T0202350

Reconsidering Causal Powers

Reconsidering Causal Powers

Historical and Conceptual Perspectives

Edited by
BENJAMIN HILL, HENRIK LAGERLUND,
and
STATHIS PSILLOS

OXFORD
UNIVERSITY PRESS

OXFORD
UNIVERSITY PRESS

Great Clarendon Street, Oxford, OX2 6DP,
United Kingdom

Oxford University Press is a department of the University of Oxford.
It furthers the University's objective of excellence in research, scholarship,
and education by publishing worldwide. Oxford is a registered trade mark of
Oxford University Press in the UK and in certain other countries

© the several contributors 2021

The moral rights of the authors have been asserted

First Edition published in 2021

Published in the United States of America by Oxford University Press
198 Madison Avenue, New York, NY 10016, United States of America

British Library Cataloguing in Publication Data
Data available

Library of Congress Control Number: 2020951914

ISBN 978-0-19-886952-8

DOI: 10.1093/oso/9780198869528.001.0001

Printed and bound by
CPI Group (UK) Ltd, Croydon, CR0 4YY

To the memory of Joseph Rotman (1935–2015)
for his unwavering support of philosophy

Contents

List of Contributors

Deborah Brown is Professor of Philosophy at the University of Queensland, Australia.

Lisa Downing is Professor of Philosophy at Ohio State University, USA.

Brian Ellis is Emeritus Professor of Philosophy at the University of La Trobe, Australia.

Benjamin Hill is Associate Professor of Philosophy at The University of Western Ontario, Canada.

Andreas Hüttemann is Professor of Philosophy at the University of Cologne, Germany.

Henrik Lagerlund is Professor of the History of Philosophy at Stockholm University, Sweden.

Jennifer McKitrick is Professor of Philosophy at the University of Nebraska, USA.

Peter Millican is Gilbert Ryle Fellow and Professor of Philosophy at Hertford College, University of Oxford, UK.

Calvin G. Normore is Professor of Philosophy at UCLA, USA.

Walter Ott is Professor of Philosophy at University of Virginia, USA

Stathis Psillos is Professor of Philosophy of Science and Metaphysics at the University of Athens, Greece.

Howard Sankey is Professor of Philosophy at Melbourne University, Australia.

List of Abbreviations

A   Aristotle. 1985. *The Complete Works of Aristotle: The Revised Oxford Translation*. Ed. Jonathan Barnes. 2 vols. Princeton: Princeton University Press.

AT   Descartes, René. 1964–76. (1897–1913) *Oeuvres de Descartes*. Ed. Charles Adam and Paul Tannery. 12 vols. Paris: J. Vrin.

B   Boyle, Robert. 1991. *Selected Philosophical Papers of Robert Boyle*. Ed. M. A. Stewart. Indianapolis: Hackett Publishing.

CSM   Descartes, René. 1984–5. *The Philosophical Writings of Descartes*. Ed. and trans. John Cottingham, Robert Stoothoff, and Dugald Murdoch. 2 vols. Cambridge: Cambridge University Press.

CSMK   Descartes, René. 1991. *The Philosophical Writings of Descartes: The Correspondence*. Ed. and trans. John Cottingham, Robert Stoothoff, Dugald Murdoch, and Anthony Kenny. Cambridge: Cambridge University Press.

DM   Suárez, Francisco. 1965 (1597). *Disputationes Metaphysicae*. Hildesheim: G. Olms.

EHU   Hume, David. 2000. *An Enquiry concerning Human Understanding: A Critical Edition*. Ed. Tom L. Beauchamp. Oxford: Clarendon Press.

Essay Locke, John. 1975. *An Essay concerning Human Understanding*. Ed. Peter H. Nidditch. Oxford: Clarendon Press.

Search Malebranche, Nicolas. 1997. *The Search After Truth with Elucidations of The Search after Truth*. Trans. Thomas M. Lennon and Paul J. Olscamp. Cambridge: Cambridge University Press.

ST   Aquinas, Thomas. 1947–8. *Summa theologica*. Trans. the Fathers of the English Dominican Province. 3 vols. New York: Benziger Brothers.

T   Hume, David. 2007. *A Treatise of Human Nature: A Critical Edition*. Ed. David Fate Norton and Mary J. Norton. 2 vols. Oxford: Clarendon Press.

THM   La Forge, Louis de. 1997. *Treatise on the Human Mind*. Trans. Desmond M. Clarke. Dordrecht: Springer.

Introduction

Benjamin Hill, Henrik Lagerlund, and Stathis Psillos

The present volume brings together twelve scholars from different philosophical disciplines (metaphysics, philosophy of science, and history of philosophy) with the aim to discuss the ontic status of causal powers, blending together historical and conceptual perspectives. Most of the authors in this volume had the privilege of discussing early drafts of our papers among ourselves and with students during a Summer Institute organized by the Rotman Institute of Philosophy—Engaging Science, of the University of Western Ontario and partly funded by an SSHRC grant. The event lasted for two weeks in July 2014, and took place in London, Ontario, and on the shores of the Lake Huron.

The motivating idea behind this detailed introduction and the twelve specially written papers is that understanding the resurgence of powers in current metaphysics and philosophy of science, as well as their role in the current scientific image of the world, is best achieved by mapping the trajectory (both conceptually and historically) of the movement *away* from powers and the dominant Aristotelian conception of nature in the seventeenth century and *towards* powers at the end of the twentieth century. This mapping will cast light on the key arguments against powers in the seventeenth century, will unravel their philosophical and scientific presuppositions, and will examine their relevance (or lack thereof) for the current debate about powers. At the same time, this mapping will locate the current trend to revitalize powers within its proper historical and conceptual context and attempt a comparison of the current conceptions of powers with the ones offered by Aristotle and the medieval Aristotelians and criticized by the *novatores*. Along the way, the attempted mapping will offer reappraisals of dominant views about, or dominant interpretations of, various important philosophers, such as Descartes, Hobbes, Malebranche, Locke, and Hume.

When all is said and done, it might appear that the elimination or reconceptualization of powers that took place in the seventeenth century has been a long(ish) interval in an otherwise continuous appeal to *sui generis* causal

Benjamin Hill, Henrik Lagerlund, and Stathis Psillos, *Introduction* In: *Reconsidering Causal Powers: Historical and Conceptual Perspectives.* Edited by: Benjamin Hill, Henrik Lagerlund, and Stathis Psillos, Oxford University Press (2021).
© Benjamin Hill, Henrik Lagerlund, and Stathis Psillos.
DOI: 10.1093/oso/9780198869528.003.0001

powers to ground and explain activity in nature. And yet, powers are subject to scrutiny and criticism today as they were in the seventeenth century and for more or less the same reasons.

In the rest of this long introduction and before a summary of the individual papers, there is an attempt (not wholly impartial) to sketch the key conceptions of, and arguments for and against, powers from Aristotle up to the present. This narrative is, of essence, selective and incomplete, but it hopefully offers the basic contours of the movement away from powers and back to them, and the backdrop against which the essays are set. The papers offer more nuanced accounts as well as alternative interpretations of parts of this narrative.

1. The History of the Powers Debate

1.1. Aristotelian Powers

Once upon a time powers ruled the world. In his *Sophist*, Plato had the Eleatic Stranger tie real existence with the power to produce change or to undertake change. Here is the characteristic quotation: 'everything which possesses any power of any kind, either to produce a change in anything or to be affected even in the least degree by the slightest cause, though it be only on one occasion, has real existence' (Plato 1997, 247d–e, 269; translation amended). To be, then, is to be powerful. Later, Aristotle (A 1046a10–11, 2:1652) developed the first systematic theory of power. A power is 'a principle of change in something else or in itself *qua* something else'. Principles are causes; hence powers are causes. Powers are posited for explanatory reasons—they are meant to explain activity in nature: change and motion. Action requires agency. For *X to act on Y*, X must have the (active) power to bring a change to Y, and Y must have the (passive) power to be changed (in the appropriate way) by X. Powers have modal force: they ground facts about necessity and possibility. Powers necessitate their effects: when a (natural) power acts (at some time and in the required way) and if there is 'contact' with the relative passive power, the effect necessarily (that is, inevitably) follows. Here is Aristotle's (324b8, 1:530) example: 'and that *that which can be hot* must be made hot, provided the heating agent is there, i.e. comes near.'

At the same time, what is possible for X to do is exactly what it has the power to do. Power (qua attribute of substances) belongs to the category of relative (*pros ti*). This is because 'their very essence includes in its nature a reference to something else' (1021a27–8, 2:1612). Positing (both active and passive) powers implies that they 'point to' (are directed towards) something outside themselves, viz. something that will suffer change when they act (having the suitable passive power) or something that will cause change in them (having the suitable active power).

Aristotelian powers are real, irreducible, inherent in substances, causal, and relative. Moreover, powers are possessed by their objects even when they are not acting. In fact, that X has the power to Φ and yet X is not Φ-ing captures *what it is* for X to have the power to Φ. Hence, what it is for X to possess the power to Φ is not the same as X actually Φ-ing (acting). X's having the power to Φ grounds and explains X's Φ-ing.

This theoretical account of powers was retained, more or less intact, throughout the Middle Ages and became the major explanatory framework for all kinds of accounts of the phenomena. Powers were taken to be natural tendencies (inclinations) that objects possess in virtue of their (God-given) nature. Hence, objects possess specific natural tendencies and lack others. These natural inclinations act of necessity (so long as the power is unimpaired) and act uniformly. Importantly, powers are directed to their acts (Aquinas also highlights this, see *ST* 1.77.3, 1:385–6.). Hence, powers 'point' to something outside themselves. Qua relatives, they have a being toward something else. They have an 'esse ad'. This towardness of powers reinforces the teleological element in nature that characterized Aristotelianism. Take gravity, viz. the downward movement of (heavy) bodies. To this, Aquinas (1a.36.1, 1:748) assigned a 'twofold cause'. The final cause of a heavy body's downward movement is the lower place: the body is seeking the lower place. The efficient cause ('principle of movement') is a natural inclination resulting from gravity.

The locus of powers of an object was its substantial form. Forms are active: the active principle of a thing is a form. As Aquinas put it, 'Heat is a disposition to the form of fire' (1a.74.4, 1:921). The form was the very nature of simple things or a constituent of the nature of a thing, which was composed of matter and form. Given the diversity of forms, there is no general and unified explanation of natural phenomena: each particular causal explanation is distinct from any other in that it appeals to an individual *sui generis* form.

The Middle Ages brought with it an important and ineliminable agent: God. He fixes the natural order by giving all creatures their natures, that is, their causal powers. This creates a network of relations of necessitation among distinct existences and hence a network of natural necessities. These could be violated by God (for example, God could make fire not burn the three boys in the fiery furnace), but if nature is left to run its own course, natural necessity rules. As John Buridan (quoted in King 2001, 16) put it, 'There is another necessity which is called "natural", which is not necessity *simpliciter*, but which would be necessity with all supernatural cases put to one side.' For Aquinas, natural necessity has an ineliminable teleological element. It is an impression from God—God directs things to their ends. God fixes the nature of things and they tend towards their ends. Natural necessity is evidence for divine providence.

There are two ways in which something may happen outside the natural order. The first is when there is action against the natural powers or tendencies of things—as when an agent throws a heavy thing upwards against its natural inclination to move downwards. This action is against nature. The second is when there is action against the natural inclinations of a thing but is caused by the agent on whom these natural inclinations depend—as when in an ebb of tide there is movement of the water against its natural inclination to go downwards. Aquinas (*ST* 1.105.6, 1:519–20), who uses this example, notes that this kind of action, dependent as it is on the influence of a heavenly body, is not against nature. God then acts on nature in the second way. Hence, if God does something outside the natural order (hence against the natural inclinations of things), this is not against nature.

1.2. The Revolt against Powers

This, in broad outline, is the power ontology against which the *novatores* of the seventeenth century revolted—a revolt that had partly started already in the fourteenth century. With the mechanical account of nature and its two catholic principles—matter and motion—in place, the critique of the medieval-Aristotelian account of powers acquires centre stage. The *novatores* put forward a different conception or explanation of natural phenomena in terms of the microscopic (and hence unobservable) parts of matter and their interactions (collisions) subject to universal laws. Matter is divested of its *sui generis* causal powers. Descartes, in a now famous passage of his posthumously published *Le monde*, states boldly:

> Someone else may if he wishes imagine the 'form' of fire, the 'quality' of heat, and the 'action' of burning to be very different things in the wood. For my own part, I am afraid of going astray if I suppose there to be in the wood anything more than what I see must necessarily be there, so I am satisfied to confine myself to conceiving the motion of its parts. For you may posit 'fire'

and 'heat' in the wood, and make it burn as much as you please: but if you do not suppose in addition that some of its parts move about and detach themselves from their neighbours, I cannot imagine it undergoing any alteration or change.   (AT 11:7; CSM 1:83)

Fire burns wood not because it has some *sui generis* burning power based on its substantial form but because its minute parts move rapidly and constantly and collide with the minute parts of wood. Motion is the principle of change and not substantial forms, which are taken to be 'occult qualities'. Indeed, Descartes wrote to Regius in January 1642, 'If they say that some action proceeds from a substantial form, it is as if they said that it proceeds from something they do not understand; which explains nothing' (AT 3:506; CSMK 208–9).

But the most puzzling feature of powers was their towardness. Descartes put forward the well-known 'little souls' argument. Only minds can have content which is about something else. Hence to attribute powers to things in virtue of which they behave is like attributing little minds to them. In a letter to Mersenne dated 26 April 1643, he noted: 'I do not suppose there are in nature any real qualities, which are attached to substances, like so many little souls to their bodies, and which are separable from them by divine power' (AT 3:648; CSMK 216). Gravity is a case in point. His medieval predecessors, Descartes thought, took gravity to be a real quality 'of which all we know is that it has the power to move the body that possesses it towards the centre of the earth' (AT 3:667; CSMK 219), thereby wrongly attributing to it soul-like attributes.

If powers can no longer play the role of necessitating principles, what plays this role? The laws of nature. Descartes, in the *Principles of Philosophy*, introduces the idea that matter and its motion are governed by universal and few-in-number laws of nature, which follow directly from God's immutability and constitute the 'the secondary and particular causes of the diverse movements which we notice in individual bodies' (AT 8a:62; CSM 1:240–1). The idea that laws can be causes of motion is so alien to the current mind that it does not seem to make sense. And yet it does (proving that Descartes's conception of laws of nature is so different from ours). The (rough) idea is this: if bits of matter are inert and powerless something must causally determine their behaviour when they move on their own and when they collide with others. Descartes's radical claim is that God does this (qua the primary and universal cause of all motions) *via* laws. Laws then are God's tools (instruments) for the redistribution of quantities of motion to bits and pieces of matter, while, due to God's immutability, the total amount of quantity of motion in the universe, set by God during the Creation, is conserved.

In the Cartesian occasionalism, God—via simple and comprehensive laws, which He chooses—is the sole cause of all motion. Father Malebranche argued that Aristotle's idea of nature, that is the idea that things have *natures* in virtue of which they behave—of necessity—the way they do is a 'pure chimera'. We have no conception of power of secondary things; no object can move itself, let alone another thing. Occasionalism emerges as the view that bodies lack motor force and God acts on nature via general laws. Not all *novatores* wanted to eliminate powers. But most took it that power has a place in the new mechanical theory *only if* it is suitably connected with the primary qualities of matter and their motion. In his famous lock and key case, Robert Boyle takes it that both the key and the lock, qua pieces of metal with certain sizes and shapes, acquire a new capacity, when by means of motion they are 'applied to one another after a certain manner' (B 23): the key unlocks the lock. But this capacity is *new* not in the sense that any new real or physical entity was added to the key or the lock. It's new in the sense that it does not supervene on either the key or the lock, but it supervenes on both and their relations. Boyle is adamant that, though his scholastic predecessors would look at this capacity as 'a peculiar faculty and power in the key that it was fitted to open and shut the lock' (23), nothing of the kind is the case. Nothing real (a new real quality) was added to each object taken separately. Nothing 'distinct from the figure it had before those keys were made' (23). But, of course, 'nothing new' does not imply nothing. Due to being part of a relational structure, the body—qua matter—acquires a modification (and not a *res*) which enables it 'to produce various effects', which it would not have been able to have had it not been part of this relational structure. Given these new effects, we 'make bodies to be endowed with qualities', 'upon whose account' the effects follow. But these qualities are not 'any real or distinct entities, or differing from the matter itself furnished with such a determinate bigness, shape, or other mechanical modifications' (24).

John Locke took it that objects have powers (active and passive). He (*Essay* II.xxi.1, 233) says: 'Fire has a power to melt Gold, i.e. to destroy the consistency of its insensible parts, and consequently its hardness, and make it fluid; and Gold has a power to be melted.' But these powers are not like scholastic faculties. Why is meat digested in stomachs? To say, like the scholastic philosophers did, that the stomach has a digestive faculty is trivial and non-explanatory. It amounts to saying, 'Digestion is performed by something that is able to digest' (II.xxi.20, 244). They are powers suitable for mechanical philosophy: they depend on the micro-constitution of objects. Like Boyle, Locke takes it that 'Power includes in it some kind of relation' (II.xxi.3, 234).

This is analysed in Draft B of the *Essay*, where Locke talks about the purgative power of rhubarb. This is grounded in the microstructure of rhubarb. It is relative nonetheless because whether or not this microstructure has a purging power depends on whether there are forms of life with the right kind of digestive system. For Locke (1990, 262) 'the purgeing power in Rhubarb is relative for rhubarb would still be the same were there noe animal in the world capable of being purged.' The ascription of power then depends on the relational structure in which rhubarb is embedded.

Significantly, Leibniz took it that extension is not enough for matter and that some notion of inherent power should be added. Based on a general metaphysical principle that 'there is neither more nor less power in an effect than there is in its cause' (Leibniz 1989, 125), he argued that a 'formal principle' must be added to 'material mass' as an irreducibly dynamical notion, which accounts for the power there is in matter. This notion of 'power' in bodies, Leibniz explains, is twofold. It is passive force when it is thought of as 'matter or mass' and active force when it is thought of as constituting 'entelechy or form.' Matter is dynamical precisely because it is characterized by impenetrability and resistance to motion (inertia). But there is active power too. Leibniz takes this to be distinct from the scholastic *potentia*. He takes the scholastic notion of power to be passive in the sense that it amounts to 'receptivity to action.' Whereas Leibniz's power is active *already* in the sense that it involves an actual *conatus* or tendency toward action such that 'unless something else impedes it, action results' (252). The notion of active power plays a role both in metaphysics *and* in physics, with the former role being the primary and the latter being the derivative. The primary sense of active power is metaphysical, since it is what makes a substance what it is—its form. This form is dynamical but at the same time mental; it is 'either a soul or a form analogous to a soul' (Leibniz 1989, 162). But if the primary sense of active power is what makes a substance what it is, its relevance to physics is nowhere to be seen unless it somehow gets connected with natural forces. Hence Leibniz talks about a derivative power, known as 'impetus, conatus, or a striving [*tendentia*] ... toward some determinate motion' (Leibniz 1989, 253).

Opposing occasionalism, Leibniz argued that powers are 'in bodies' and they cause (or causally contribute) to the motions of bodies. Though for Leibniz the motion of matter is law-governed, where the laws are instituted by God on the basis of maximizing simplicity and perfection (comprehensiveness), bodies have innate forces by virtue of which they act according to laws.

Perhaps surprisingly, in his unpublished text *De gravitatione*, Isaac Newton took a rather hard line on powers. Criticizing Descartes's account of matter as extension, Newton (2004, 21) argued that, though it is not customary to 'define substance as an entity that can act upon things, yet everyone tacitly understands this of substances'. He then went on to define a body as something with three kinds of power: impenetrability, mobility, and capacity to affect our senses. To be sure, these are conditions imposed by God on determined quantities of extension. But they are powers nonetheless. Newton doesn't say much about their ontic status. But he does say two important things. First, they are not like the scholastic substantial form, and second, they are subject to laws. In fact, the key power of bodies to move amounts to the transference of impenetrability from one part of space to another 'according to certain laws' (28). Besides, 'we do not move our bodies by a proper and independent power but by laws imposed on us by God' (30).

This idea of law-constituted powers Newton did not develop further in *Principia* except in a crucial respect when it comes to gravity. In his famous 'General Scholium' from *Principia*, Newton declared that he feigns no hypotheses for the cause of gravity. He added rather boldly that hypotheses have no role to play in experimental philosophy. And he stressed:

> In this experimental philosophy, propositions are deduced from the phenomena and are made general by induction. The impenetrability, mobility, and impetus of bodies, and the laws of motion and the law of gravity have been found by this method. And it is enough that gravity really exists and acts according to the laws that we have set forth and is sufficient to explain all the motions of the heavenly bodies and of our sea. (1999, 589)

What's important for our purposes is Newton's claim that gravity exists and acts according to *laws*. The inverse square law constitutes gravity, at least when it comes to its causal-explanatory role. That there may be an unknown causal basis for this power is irrelevant. In an unsent letter written circa May 1712 to the editor of the *Memoirs of Literature*, Newton stressed that it is not necessary for the introduction of a power—such as gravity—to specify anything other than the law it obeys; no extra requirements should be imposed, and in particular no requirement for a mechanical grounding. He wrote:

> And therefore if any man should say that bodies attract one another by a power whose cause is unknown to us, or by a power seated in the frame of nature by the will of God, or by a power seated in a substance in which bodies move and float without resistance and which has therefore no *vis inertiae* but acts by other laws than those that are mechanical: I know not why he

should be said to introduce miracles and occult qualities and fictions into the world.   (2004, 116)

In 1713, in the preface to the second edition of *Principia*, Roger Cotes noted emphatically that gravity is a 'primary affection of bodies' and that, like other primary affections such as extension, mobility, and impenetrability, is one of the primary qualities of all bodies universally (Newton 1999, 391). Indeed, Newton employed the famous third rule of philosophy, which is a rule of induction, in order to argue that 'all bodies gravitate toward one another'. And yet he added that he is not affirming that 'gravity is essential to bodies' (796).

In a 1693 letter to Richard Bentley, Newton (2004, 102) had stressed that it is 'inconceivable that inanimate brute matter should, without the mediation of something else which is not material, operate upon and affect other matter without mutual contact, as it must be, if gravitation in the sense of Epicurus, be essential and inherent in it'. It is an absurdity, he later added in print, that 'gravity should be innate, inherent, and essential to matter, so that one body may act upon another at a distance through a vacuum, without the mediation of anything else, by and through which their action and force may be conveyed from one to another' (1999, 796). The only inherent force is 'the force of inertia'. Newton took it that, if gravity were essential to bodies, then there would be action at a distance. If action at a distance is to be avoided, gravity should not be taken to be essential to bodies, qua material substances.

But this does not imply that gravity is not a universal quality of bodies. Gravity depends on the mass of bodies, which is an essential and universal quality of bodies. And Newton has shown how exactly gravity depends on mass in the law of universal attraction. In the 'General Scholium' Newton sums up his findings about gravity thus:

> Thus far I have explained the phenomena of the heavens and of our sea by the force of gravity, but I have not yet assigned a cause to gravity. Indeed, this force arises from some cause that penetrates as far as the centers of the sun and planets without any diminution of its power to act, and that acts not in proportion to the quantity of the surfaces of the particles on which it acts (as mechanical causes are wont to do) but in proportion to the quantity of solid matter, and whose action is extended everywhere to immense distances, always decreasing as the squares of the distances.   (1999, 943)

If the action is extended everywhere to immense distances and it obeys a law, gravity is hardly a non-universal force. In fact, gravity exists insofar as it acts according to laws. This is the law-constitutive approach to powers introduced by Newton.

1.3. Hume's Critique of Powers

By the time Hume's *Treatise* appeared, *sui generis* causal powers were far from abandoned. This made Hume offer a systematic critique of the ontic and epistemic status of powers. Echoing Malebranche, Hume put forward a disagreement argument: there is a diversity of philosophical views about how powers play their role and 'none of them has any solidity or evidence'. The whole panoply of substantial forms, accidents, and faculties is perfectly unintelligible and inexplicable (*T* 158). Then, in line with his overall empiricist approach which implies that legitimately held ideas require prior impressions, he advanced a verificationist argument: there is no impression of efficacy or power; 'we never have an impression that contains any power or efficacy' (160). Hume argued that ''tis impossible we can have any idea of power or efficacy, unless some instances can be produc'd, wherein this power *is perceiv'd* to exert itself.' Hence, there cannot be unmanifested powers, that is, powers which exist, even though there are no impressions of their manifestations. In fact, the very ascription of a power to a thing requires that this power has been exercised. Hence, 'The distinction, which we often make betwixt *power* and the *exercise* of it, is equally without foundation' (171).

To all this he added what might be called the 'unnecessary duplication argument': power or efficacy is not to be confused with an unknown quality of the objects—it is taken to be signify something that grounds the action and causation of bodies. But the constant conjunction of ordinary qualities of the bodies is enough to account for action and causation as well as for the idea of efficiency (171).

Finally, Hume advanced the key argument that powers cannot be necessity-makers. Powers were supposed to ground the necessary connection between cause and effect. But what kind of necessity is this? Hume takes it that if there were such a necessity, it would be such that the effect would be entailed by the cause—and the entailment would amount to a demonstration of the effect on the basis of the cause—hence the impossibility of the cause without the effect. But it is conceivable that the cause is unaccompanied by the effect and conversely. Hence, there is no metaphysical necessity in causation.

Could it be some weaker kind of necessity—natural necessity—which characterizes the power-based connection among distinct existences? In line with the standard distinction between the absolute and the ordained power of God, there are philosophers who attribute efficacy to secondary causes (natural bodies) and hence a 'derivative, but a real power and energy to matter' (*T* 161). It could then be argued that, though it is metaphysically possible that though A has the power to produce B, A exists without B, it is naturally necessary. Hume's far-reaching point was precisely that this appeal to natural necessity can be questioned.

For Hume this kind of issue takes the following form: can there be knowledge of Principle of Uniformity of Nature (PUN) based on experience, and in particular on probable reasoning? His famous answer is that this kind of move would be question-begging. Probable reasoning begins with experience and memory (89–90) and should provide a connection between something given in experience and something not given in experience. Hence, probable reasoning is causal inference: the only 'just inference from one object to another'. But then here is the problem: probable inference of the sort required, being causal inference, is founded on the 'presumption of a resemblance betwixt those objects, of which we have had experience, and those, of which we have had none'. Hence, this presumption (PUN) cannot come from probable reasoning. The very same principle cannot be the ground for an inference *and* the product of this inference. Or, as Hume put it, 'The same principle cannot be both the cause and effect of another; and this is, perhaps, the only proposition concerning that relation, which is either intuitively or demonstratively certain' (90).

Could it be that powers are posited to ground causal inference? This might well be a legitimate metaphysical inflation of the world by means of powers. Here is the relevant argument, as summed up by Hume: 'The past production implies a power: The power implies a new production: And the new production is what we infer from the power and the past production' (90). Hence, the argument for powers might be the following:

1. A has been constantly conjoined with B.
2. A has the power to produce B.
3. A necessarily produces B.
4. Hence All As are B.

It's no surprise that Hume argues that this argument begs the question, since it relies on a power-based PUN. The move from 2 to 3 and then to 4 requires that 'the same power continues united with the same object, and that like objects are endow'd with like powers' (91).

Suppose that powers were allowed in the sense that the production of an object B by object A implies a power in A. It is still the case that the past exercisings of A's power to produce B are not enough to prove that A necessarily has the power to cause B, that is A causes B *simpliciter*. The reason is that past experience 'can only prove, that that very object, which produc'd any other, was at that very instant endow'd with such a power; but can never prove, that the same power must continue in the same object or collection of sensible qualities; much less, that a like power is always conjoin'd with like sensible qualities.' So, unless a variant of PUN is assumed, the conclusion that A causes B does not follow. But this PUN-like principle raises the further question of its own provenance, and the standard Humean objection is repeated: 'why from this experience we form any conclusion beyond those past instances, of which we have had experience'? This would lead to regress 'which clearly proves, that the fore going reasoning had no just foundation' (91). Hume's point, then, is that powers cannot be necessity-makers either on metaphysical or epistemic grounds.

1.4. From Powers to Laws

Hume's critique of powers did not persuade Reid to stop positing powers. He spoke freely of active powers and took it that:

(a) the very concept of power is simple and undefinable;
(b) power is *not* something we either perceive via the senses or we are aware of in our consciousness (we are conscious only of the *operation* of power and not of the power itself);
(c) power is something whose existence we infer by means of reason based on its operation/manifestation;
(d) power is distinct from its exertion in that there may be unexerted powers;
(e) the idea we have of power is relative, viz. as the conception of something that produces or brings about certain effects;
(f) power always requires a subject to which it belongs: it is always the power *of* something; the power that something *has*; and
(g) causation is the production of change by the exercise of power.

Exercising a power requires agency and since there is no agency in nature, strictly speaking, there is no causation in nature. Indeed, Reid (2011, I.6, 41–7) insisted that, properly understood, active powers require subjects that have intelligence and will to exercise them. Inanimate matter then can be no such subject. Only God—who is an 'off-stage agent'—can be the cause possibly by means of secondary causes.

But Reid wrote after Newton and thought that the universe was law-governed. Yet, subsumption under laws of nature does not constitute causation; nor does it amount to causal explanation. For him, 'the laws of nature are the rules according to which the effects are produced; but there must be a cause which operates according to these rules. The rules of navigation never steered a ship. The rules of architecture never built a house' (I.6, 47) Hence, a cause is something that has the power to bring about an effect *in accordance with the law*; but knowing the laws does not amount to knowing the causes. From all this, he drew the rather pessimistic conclusion that, in spite of the fact that scientists have discovered a number of laws of nature, 'they have never discovered the efficient cause of any one phenomenon' (I.6, 47). Which, for Reid, is just as well since those scientists who understand what science is about and what the laws of nature are do not claim that science discovers (or aims to discover) causes. For Reid causation is tied to agency and laws of nature are not agents. As he (IV.3, 288) put it, laws of nature 'are not endowed with active power, and therefore cannot be causes in the proper sense. They are only the rules according to which the unknown cause acts.'

John Stuart Mill's critique of *sui generis* powers (or qualities, or virtues) consolidated the move from powers to laws of nature. Besides, Mill was among the first to secularize the conception of natural laws. Mill's key argument against powers is that they are the products of a double vision. Echoing Hume, but not referring to him directly, Mill argued that there is nothing more to the power of an object to produce Φ than actually producing Φ. So, when, for instance, snow is endowed with whiteness, being the power to produce in us a certain sensation of white, what is really asserted is that 'the presence of snow produces in us that sensation'. Positing the power ('something in virtue of which the object produces the sensation') is redundant. As Mill noted, this hypostatization of powers (or qualities) 'Moliere so happily ridiculed when he made one of his pedantic physicians account for the fact that opium produces sleep by the maxim, Because it has a soporific virtue' (Mill 1911, 58). According to Mill, stating the 'soporific virtue' 'merely asserted over again, the fact that it produces sleep'. More generally, when it is asserted that object A has the power P to produce Φ, it is asserted that A produces Φ.

With this 'reductive' understanding of powers in mind, Mill had no problem with talking about powers, as required by common opinion and common language, noting at the same time that this does not imply any recognition of powers 'as real existences', since this recognition is not 'warranted by a sound philosophy' (65).

As is well known, Mill tied causation to regularity; in particular, to the presence of an invariable and unconditional sequence of events. To say, then, that C causes E is not to say that C has a power to produce E. Rather it is to say that C and E are in a relation of coexistence and succession and that events like C are invariably and unconditionally followed by events like E. As he put it, 'the cause of a phenomenon' is 'the antecedent, or the concurrence of antecedents, on which it is invariably and unconditionally consequent' (245).

The condition of unconditionality is meant to exclude cases in which C and E are effects of a common cause (like the succession of night and day). Causes then are not powers but (antecedent) conditions upon which the effect (consequent) follows. Similarly, the distinction between the agent and patient vanishes. 'Patients are always agents' and 'all the positive conditions of a phenomenon are alike agents' (243). Mill attributes the search for power in nature in the 'disposition of mankind in general, not to be satisfied with knowing that one fact is invariably antecedent and another consequent, but to look out for something which may seem to explain their being so' (261). It is precisely this disposition that should be resisted. Power-ascription reduces to causation, which is uniformity.

Causation, for Mill, is tied to laws. Indeed, causation is subject to an overarching law, viz. that every fact in nature invariably succeeds some other fact which has preceded it (236). These invariable successions ('the uniformities which exist among natural phenomena') are the laws of nature, 'when reduced to their simplest expression' (231). There is a seamless transition from causation to laws and conversely, via the idea of natural uniformity. Significantly, Mill frees the concept of laws of nature from any theological connotations and underpinnings. In a beautiful passage of the *System of Logic* (230) he states that 'the expression, law of nature has generally been employed with a sort of tacit reference to the original sense of the word law, namely, the expression of the will of a superior.' Then, Mill adds, when 'it appeared that any of the uniformities which were observed in nature, would result spontaneously from certain other uniformities, no separate act of creative will being supposed necessary for the production of the derivative uniformities, these have not usually been spoken of as laws of nature.' But instead of giving up the expression 'laws of nature' as applicable to ungrounded uniformities, Mill makes his famous proposal: 'According to another mode of expression, the question, What are the laws of nature? may be stated thus:—What are the fewest and simplest assumptions, which being granted, the whole existing order of nature would result? Another mode of stating it would be thus: What are the fewest general propositions from which all the uniformities which exist in the universe might be deductively inferred?' Hence, according to Mill, the laws of nature are those regularities which are captured by axioms and theorems in a simple and comprehensive deductive systematization of our knowledge of the world. This view, adopted by Frank Ramsey (1978) and David Lewis (1973) in the twentieth century, has become the main empiricist account of laws of nature.

1.5. Bringing Powers 'Down to Earth'

The legacy of this movement of thought to the twentieth century was that powers became problematic and, ultimately, redundant. In the hands of most philosophers at least until the 1970s, powers (aka dispositions or dispositional properties or capacities) were eliminable and dispensable.

Rudolf Carnap, in his 'Testability and Meaning' (1936; 1937), set the agenda for what was to follow by suggesting that dispositional predicates such as 'visible', 'edible', 'fragile', and 'soluble' can and should be analysed by means of conditional statements. An attempt to *explicitly define* predicates such as 'is soluble in water' by reference to stimulus conditions and response conditions failed patently. An explicit definition has the form:

$$\forall x \, (Qx \longleftrightarrow (Sx \rightarrow Rx)) \quad (D).$$

(D) says that the predicate Q applies to x if and only if, when x satisfies the *stimulus-condition* S, x displays the (observable) *response* R. For instance, an explicit definition of the dispositional predicate 'soluble in water' would be like this: an object A is soluble in water if and only if if A is put into water, then A dissolves. But if the conditional $(Sx \rightarrow Rx)$ is understood as material implication, then $(Sx \rightarrow Rx)$ is true even if the stimulus-conditions S do *not* occur. So, explicit definitions turn out to be empty. To avoid this, explicit definitions must be understood as a subjunctive statement, that is, as stating that the object A has the property Q iff if A *were* to be subjected to stimulus-conditions S, then A *would* manifest the characteristic response R. All this brought in tow the problem of the truth conditions of subjunctive and counterfactual conditionals. To avoid it Carnap suggested that dispositional predicates should be introduced by reduction sentences. Unlike explicit definitions, reduction sentences do not fully define (and hence eliminate) the *definiendum*.

A dispositional predicate Q is introduced by means of the following *reduction pair*:

$$\forall x \, (S_1 x \rightarrow (O_1 x \rightarrow Qx)) \quad (G1)$$

$$\forall x \, (S_2 x \rightarrow (O_2 x \rightarrow \neg Qx)) \quad (G2) \quad (RP)$$

where S_1, S_2 describe stimulus conditions and O_1, O_2 describe characteristic responses. In the case that $S_1 = S_2$ (=S) and $O_1 = \neg O_2$ (=O), the reduction pair (RP) becomes a *bilateral reduction sentence*:

$$\forall x \, (Sx \rightarrow (Qx \longleftrightarrow Ox)) \quad (RS)$$

So, suppose that we want to introduce the predicate Q, 'soluble', by means of a reductive sentence. This will be: if the stimulus-conditions S obtain (that is, if we put object A into water), then A is soluble iff the characteristic response R obtains (that is, if A dissolves in water). This is fine as far as it goes, but it does not go very far. The reduction sentence (RS) does not fully define the predicate Q. The meaning of Q is *not* completely specified by virtue of stimulus-response conditions. At best, the reduction sentence gives, as Carnap (1936, 443) put it, 'a conditional definition' of Q. Even if the content of a dispositional predicate can be further narrowed down by the introduction of more reduction sentences, it will never be fully specified by them. Hence dispositional predicates have excess content over their conditions of application.

This excess content might well be a dispositional property (a *power*). But to many, admitting *sui generis* dispositional properties was an anathema. For one, it seems they have had a weird ontic status: they existed even if they did not act, which was reminiscent of the Aristotelian potential versus actual distinction. Hence, dispositional properties were taken to be non-actual, non-occurrent properties, which exist potentially. They were contrasted with kosher categorical properties (like shape, size, and figure) which were taken to be occurrent, actual properties of their bearers. For another, dispositional properties were associated with stimulus and response conditionals (to be fragile is to be such that if it is suitably struck it breaks), which to many meant that dispositions have a *conditional* existence.

This kind of hostility to powers was captured very powerfully by Nelson Goodman (1983, 40) who said that ascribing dispositions to a thing is like making it 'full of threats and promises'. The dispositions of a thing 'strike us... as rather ethereal'. Hence, Goodman concluded, 'we are moved to inquire whether we can bring them down to earth; whether, that is, we can explain disposition-terms without any reference to occult powers.' The spectre of occult qualities, of power and substantial forms, loomed large again. It had to be put to rest.

For Goodman the key problem with dispositional predicates is that they are modal: they apply to things in virtue of possible situations (for example, a fine china is fragile even if it never breaks). Hence, he thought that a genuine solution to the problem of dispositional predicates is to specify their ascription conditions 'in terms of manifest predicates' (42). As Carnap's work has shown, this is easier said than done. It turns out that any attempt to link dispositions with conditionals (the so-called conditional analysis of dispositions) relies on what would have happened under certain circumstances and not on what actually happens.

Be that as it may, what came to be known as simple conditional analysis states the following: something X is disposed at time t to give response R to stimulus S iff, if X were to undergo stimulus S at time t, X would give response R. Irrespective of how we understand subjunctive conditionals, this account fails because dispositions might well fail to entail any conditionals. For instance, it might be that the conditions under which a disposition is triggered are such that they cause the loss of the disposition. This is the case of the so-called finkish dispositions (see Martin 1994). For instance, think of a device (an 'electro-fink') attached to a wire such that the wire is live if and only is it is in contact with the conductor. In effect, the electro-fink makes sure that no current flows through the wire when it does not touch the conductor. Now, the conditional 'If the wire were touched by a conductor then the wire would be live' is true even though the wire is dead. But it may also be the case that the associated conditional is false because the disposition is 'masked': the disposition is not lost and yet its manifestation is prevented. Throwing a glass wrapped up in packing material is a case in point; the glass is still fragile and yet it is not the case that if it were dropped it would break. (Antidotes make a similar case.)

Though examples such as the above are meant to dissociate ascription of dispositional properties with conditionals, and hence to show that dispositions are not reducible to conditionals, the immediate response is that the simple conditional analysis is too simple to be credible. It is typically replaced by a 'reformed conditional analysis' suggested by David Lewis (1997, 157): 'Something x is disposed at time t to give response r to stimulus s iff, for some intrinsic property B that x has at t, for some time t′ after t, if x were to undergo stimulus s at time t and retain property B until t′, s and x's having of B would jointly be an x-complete cause of x's giving response r.' Let's not dwell much on this reformed analysis. Instead, let us bring out its main feature and the concomitant philosophical issue. It ties dispositions to the existence of stable intrinsic categorical properties, which objects possess as long as they possess the relevant disposition. The key idea here is that dispositions need grounding in 'categorical' properties. For instance, salt is soluble is water in virtue of its molecular structure, which is meant to be a non-dispositional property. Quine (1974, 11) went as far as to argue that the dispositional idiom is a temporary heuristic device which will have 'no place' in 'ideal language for a finished theory of reality.' Until then, ascription of a disposition is reference to a hypothetical unknown mechanism, which brings about the phenomena of interest. When these mechanisms become known, the dispositional idiom becomes redundant.

John Mackie (1977, 366), echoing Hume and Mill, argued that the idea that there are intrinsic powers—the referents of the dispositional predicates—is the product 'of metaphysical double vision'. The power is just the causal process doubled as something latent in the objects that engage in this process: water, apart from its molecular structure, has the power to dissolve salt and so on. Powers and dispositions, on this view, are 'shadows' cast by the interactions of the categorical properties of things.

1.6. Powers: the Comeback

In the last few decades of the twentieth century powers have made an impressive comeback. One of the first to take power talk at face value was Rom Harré. According to him (1970, 85), the ascription of a power to a particular has the following form: X has the power to $\Phi=_{def}$ if X is subject to stimuli or conditions of an appropriate kind, then X will do Φ, *in virtue of its intrinsic nature*. Hence, power-ascriptions are analysed in terms of:

a) a specific conditional (which says what object X will or can do under certain circumstances and in the presence of a certain stimulus); and
b) an unspecific categorical claim about the nature of X.

As Harré and Madden put it, 'To ascribe a power to a thing or material is to say something specific about what it *will* or *can do*, but to say something unspecific about what it *is*. That is, to ascribe a power to a thing asserts only that it can do what it does in virtue of its nature, whatever that is' (1975, 87). The claim about the nature of X is unspecific, because the exact specification of the nature or constitution of X in virtue of which it has the power to Φ is left open. Discovering this is a matter of empirical investigation. This, however, implies that ascription of powers is, at best, explanatorily incomplete and, at worst, trivial, unless something specific is (or can be) said about the *nature* of the thing that has the power. Otherwise, power-ascription merely states what needs to be explained, viz. that causes produce their effects. As Harré (1970, 276) admitted, 'In the ascription of powers the categorical component is like a promissory note, we only believe that it is not logically impossible for it to be cashed.' This is where generative mechanisms come in. Specifying the generative mechanism is cashing the promissory note. As he put it, 'Giving a mechanism... is... partly to describe the nature and constitution of the things involved which makes clear to us what mechanisms have been brought into operation' (124).

So, the key idea is this. Causes produce their effects because they have the power to do so, where this power is grounded in the mechanism that connects the cause and the effect and the mechanism is grounded in the nature of the thing that does the causing. A similar view was advanced in the 1980s by Nancy Cartwright (1983; 1989), who took it that capacities are irreducible, primary, and causally active constituents of the world.

In the last few decades, there has been the rise of neo-Aristotelianism. There are two main entry points of powers: the nature of properties and the nature and necessity of laws.

1.7. The Nature of Properties

Powers have reappeared as modally laden properties. Dispositionalists take properties to be (active and passive) powers, while categoricalists take properties to be purely qualitative and inert, trying to account for the presence of activity in nature on the basis of laws of nature. Dispositionalists argue that the identity of properties qua powers is given by their causal role. This view was defended by Sydney Shoemaker (1980). He suggested that all properties are best understood as powers since the only way to identify them is via their causal role (the things they can do). Accordingly, two seemingly distinct properties that have exactly the same powers (that is, causal role) are, in fact, one and the same property. And similarly, one cannot ascribe different powers to a property without changing this property. For the pandispositionalist, as Mumford and Anjum (2011, 45–6) put it, 'all things have properties and all properties are powers'.

A key argument for pandispositionalism is that it avoids quidditism. A quiddity is an individuating factor that makes a property what it is independently of its causal role. Friends of powers argue that, if properties have quiddities, these will end up being unknowable. To avoid this unwanted consequence, pandispositionalists take it that properties do not have quiddities, that is, properties have no intrinsic nature over and above their causal profile. This view has become known as causal structuralism. The main idea is that causally isomorphic structures are identical. Suppose that there are two possible worlds W_1 and W_2 such that they differ only in the following: in W_1 two properties A and B act in tandem to generate a certain causal profile Q, whereas in W_2 a *single* property C has the causal profile Q. Suppose, further, that in W_1 A or B, taken individually, do not have any further causal role. Causal structuralism entails that, all else being equal, W_1 is identical to W_2. Quidditism entails that they are different possible worlds. The argument would be that, even though we may never be able to figure out whether we live in W_1 or W_2, to make sense of this metaphysical possibility we need to go beyond causal roles. But causal structuralism would deny that this scenario is even metaphysically possible.

A key argument against pandispositionalism is the so-called 'always packing, never travelling' argument (Molnar 2003, 173; Armstrong 1997, 80). The idea is this: if properties are nothing but powers, then when a power is manifested, its *effect* (the acquiring of a property by a particular) will also be a *power*. Hence, nothing really happens apart from the shifting around of powers from particular to particular. As Armstrong (1997, 80) put it, 'Given purely dispositionalist accounts of properties, particulars would seem to be always re-packing their bags as they change properties, yet never taking a journey from potency to act.'

The friends of dispositions take it for granted that powers can be causes. Not everyone agrees. To many, dispositions are causally inert. The causing is done by the categorical property that 'realizes' or 'grounds' the disposition. Take the property of being fragile. It's not fragility that causes the vase to break (when hit hard with a hammer, let's say). Rather it is the molecular structure of the vase that is the cause of the breaking. Fragility, it is said, is a second-order property that is realized by a categorical property., viz. the molecular structure. In this setting, Prior, Pargeter, and Jackson (1982) produced a general argument for the claim that dispositions are causally impotent. It is this. Dispositions are distinct from their causal bases (since they are properties of them). But they must have a causal basis (a realizer). This causal basis (some properties) are themselves a sufficient set of properties for the causal explanation of the manifestation of the disposition (whenever it is manifested). Hence, the disposition qua distinct property is causally impotent.

A typical reply is that the grounding is done by other dispositions. But then, is it dispositions all the way down, as it were? The so-called ultimate argument for powers offers a qualified positive answer. One might acknowledge that most garden-variety dispositions are grounded in the properties of the components or parts of the entities that bear them (where these properties might be dispositional or categorical). And yet, states the ultimate argument, the fundamental properties (the properties of the fundamental particles) *are* pure or bare powers. Physics, it is claimed, posits irreducible powers: mass, charge, and spin. The fundamental particles are *simple*: they have no internal structure. Hence, they have no parts (components) which can be deemed the bearers of further properties (be they powers or non-powers) which, in turn, ground the properties of the particles. This is an empirical argument and it is very popular among the friends of powers.

How compelling is this argument? It turns out that the empirical situation might well be much more complicated than the friends of powers typically presume. The fundamental properties of the elementary particles are determined by global and local symmetries that exist in nature. More specifically, they emerge as invariances under sets of (global or local) transformations, which form a group. So, the properties of elementary particles are far from being ungrounded, though they are grounded in symmetries. The role of symmetries in the metaphysics of nature has yet to be fully understood. But no matter what one thinks on this issue, symmetries do identify fundamental properties in a way that does *not* involve what they can do.

1.8. The Nature of Laws of Nature

A second entry point for powers is the status and necessity of laws of nature. During most of the twentieth century, philosophers of science and metaphysicians took it that a broadly empiricist-Humean conception of laws of nature is the right one: Laws are universal regularities. Hence laws are contingent. A key problem with this view was the following: how can laws be distinguished from accidentally true generalizations? The answer was, typically, that something should be added to a regularity to count as a law—something robust enough to distinguish it from accidents but thin enough not to sneak in any kind of necessity. The prevailing view was that of Mill (1911), as revamped by Ramsey (1978) and Lewis (1973): the regularities that constitute the laws of nature are those that are expressed by the axioms and theorems of an ideal deductive system of our knowledge of the world, and in particular, of a deductive system that strikes the *best* balance between simplicity and strength. Simplicity is required because it disallows extraneous elements from the system of laws. Strength is required because the deductive system should be as informative as possible about the laws that hold in the world. Whatever regularity is not part of this *best system* is merely accidental: it fails to be a genuine law of nature. This approach came to be known as MRL (Mill–Ramsey–Lewis) view.

But many philosophers felt that this view does not offer a robust and objective answer to the question: what is a law of nature? A particularly sticky problem was the modal force of laws: laws don't merely state what things do, but also what they must (or cannot) do. A first reaction to the MRL was by David Armstrong (1983), Fred Dretske (1977), and Michael Tooley (1977) (the ADT approach), which took it that the laws are not the regularities themselves, though they entail the regularities. On this approach, lawhood is a certain contingent necessitating relation among properties (universals). Accordingly, it is a law that all Fs are Gs if and only if there is a relation of nomic necessitation N(F, G) between the universals F-ness and G-ness such that all Fs are Gs. Given that N is a *contingent* relation of necessitation, it's a worldly add-on; it might fail to obtain in other possible worlds.

Despite its attractions, this view faced two important problems. One is that it is unclear, to say the least, why N(F, G) entails all Fs are Gs. The motivation for the ADT approach was that there must be something that makes a law (viz. a regularity) *law*—otherwise laws float on nothing. It turns out that, if this something is this mysterious N (the *contingent* relation of necessitation), it fails to account for the existence of regularity. By the same token, (problem number 2) N, being contingent, turns out to be itself a free-floater. There is no explanation of why N obtains. It is just a brute fact that certain universals (and not others) are related by N.

As a reaction to this second problem, it was suggested that, far from being an external (contingent) relation of necessitation N, there is an internal relation of necessitation, that is, a relation that bilaterally supervenes on the intrinsic properties of the relata. Given that the relata (the related properties) are what they are, they must be so related—that is, there has to be a necessitating relation between them. On this view, the necessitating relation between distinct properties F and G is such that it is (metaphysically) impossible that F is instantiated without G being instantiated. But for properties to be able to stand in non-trivial relations to each other, they must be *powers*. Only then can properties be such that, though distinct existents, they can nonetheless bear non-trivial relations to each other (in particular, relations of entailment and 'repugnance,' as they would put it in the Middle Ages). On this view, it does not even make sense to say that properties are united by laws. Rather, properties—qua powers—*ground* the laws. Laws of nature end up being metaphysically necessary.

But here there is a split between realist lawlessness, advocated by Mumford (2004), and realist lawfulness, advocated by Bird (2007) and Ellis(2001). On the first view, precisely because properties are modally loaded, that is they are powers, and, as such, they bear certain relations to each other, viz. relations of necessary connection, exclusion, and production, there are no laws. Whatever job laws were supposed to do, it is done by powers. Hence, laws are not real (see Mumford 2004, 193). Other dispositionalists (Bird 2007 and Ellis 2001), however, argue that laws are real but not fundamental: they supervene on powers.

1.9. Dispositional Essentialism

The two entry points for powers (identity of properties and status and necessity of laws) are wrought together in dispositional essentialism. This view, defended by Alexander Bird (2007) and Brian Ellis (2001) among others, puts together two metaphysical theses:

(A) Properties have essences—that is, certain characteristics that a property has in virtue of which it is the property it is.
(B) Properties are powers—they are individuated by their causal profile, that is, by their relations they stand to other properties and the effects they produce under certain circumstances.

A and B imply that the essences of properties are powers.

Not all dispositionalists are essentialists. Some (for example, Mumford 2004) take it that all elements of the causal profile of a property are on a par. Not all dispositionalists are pandispositionalists. Some, for example Ellis (2001), take it that there are structural-categorical properties. What all neo-Aristotelians share in common is the claim that powers are irreducible, primary, and causally active constituents of the world: powers are taken to be *intrinsic and actual properties whose nature is to be directed towards their characteristic manifestation.* According to Molnar (2003), intentionality is not *solely* the mark of the mental. Physical powers too, he argued, are intentional properties: they are *directed* towards their (possibly non-existent) manifestation. This is a key feature of powers that distinguishes them from non-powers. It goes back to Aristotle and Aquinas. What the early modern philosophers thought was a mysterious feature of powers, and a reason to reject them, has become a standard feature of powers, and a reason to accept them.

2. Summary of Contributions

These, in outline, are some of the major phases and arguments of the movement away from powers and towards them. Let us now present the main points made and issues raised by the individual chapters.

2.1. 'The Inherence and Directedness of Powers', by Stathis Psillos

Stathis Psillos explores a fundamental ontological puzzle pertaining to powers, in our opinion perhaps the most fundamental one, namely the internal coherence of their dual nature. On the one hand, powers are intrinsic properties of their objects. They are really and truly predicated of their subjects because they are, strictly speaking, constituents of their subjects. Thus their beings are tied to their objects in a way that makes the rest of the world irrelevant for them, and this way is typically characterized through a counterfactual: even if everything else in the world were different, or if nothing else existed in the world, object X would possess any and all powers Φ that it now possesses in the actual world. But, on the other hand, powers are also directed toward their manifestations. This notion of directed is somewhat vague and difficult to nail down, but at the very least it prima facie implies that powers in some important way depend on something external to it and its object. The implication and its externality are crystal clear in most cases: an active/passive power requires the existence of its co-relative as a necessary condition for that power's ever being manifested. And insofar as the existence of a power depends on the possibility of its being manifested, the existence of any power Φ now seems to require the existence of its co-relatives and, consequently, the bearers of those co-relatives. Psillos begins his exploration from this prima facie conceptual tension, 'How can a power be both intrinsic to its bearer and "have its being" toward something outside of it?' (47).

Psillos traces the origins of this puzzle back to the conceptual structure of Aristotle's original account of powers, even though Aristotle himself seemed insensitive to it. Aristotelian 'powers', Psillos points out, 'have two modes of being: in potency and in act' (54). This opens up the possibility of distinguishing between a power's possession by an object and its manifestation in a causal event. The ontological duality that generates Psillos's puzzle is then laid on top of this potency-act distinction; it arises through the directedness of powers, a directed they have regardless of whether they obtain in potency or in act: 'but either way, being in potency and being in act, powers are toward something' (54). Aristotle himself did not see this notion of directedness as noteworthy, Psillos emphases, but in the Aristotelian revival of the High Middle Ages, 'it becomes central in understanding powers' (56). Psillos then turns from Aristotle to explore Aquinas's developments of this feature of directedness.

Aquinas followed Aristotle in holding that powers were qualities and relatives simultaneously. And as in Aristotle, in Aquinas this ontological or categorical duality was what 'grounds their two main features: inherence and directedness' (59). But Aquinas, unlike Aristotle, worried about how to handle cases in which relatives are lost or acquired even though the subject did not change in any qualitive way, that is the subject remained the same intrinsically. Aquinas appealed to the analogy of a root—the relative was first in the subject as a root—which Psillos unpacks as follows: 'When the power is actualized it enters into actual and real relations with the correlative powers. But because the power exists potentially too, the actualization (and hence the real relation) was in a sense in the substance "in some way as in a root". When a power shifts from potentially to actuality there is a real relation "inasmuch as one thing receives something from another or confers it upon the other"' (62). Psillos then hooks this characterization of the root analogy to Francisco Suárez's distinction between *inesse* and *esse-ad* to ground the Thomistic account in a bifurcated account of being (62–64). But the difference between *inesse* and *esse-ad*, argues Psillos, is only conceptual and not real, so this provides a coherent and feasible resolution to the puzzle of directedness that looks a lot like the 'so-called "identity theory" of properties' defended by John Heil and C. B. Martin (65). Psillos concludes:

> Powers, then, are real, but not really distinct from their qualitative grounds. Their directedness is the potential by which, qua relatives, powers endow their bearers to enter into relations with other substances. If there were a lonely object in the world, the potential for relations would be there (since the object would have some inherent qualitative properties) but no relatedness; and hence no change. But since there are various entities (substances)

in the world, powers are actualized and this accounts for the natural order in the world, viz. for substances being related to each other.   (66)

Whether contemporary friends of powers would want to embrace this inflated, bifurcated account of being, and whether they should, are questions that Psillos leaves for the reader to explore.

2.2. 'Powers, Possibilities, and Time: Notes for a Programme', by Calvin G. Normore

Of equally fundamental importance to the current debate over causal powers are its Megaric consequences, the connection between powers and modality. One of the central motivations for adopting a powers ontology is said to be the support causal powers provide for grounding and explaining alternative possibilities. Calvin Normore provides a robust defence of this idea by defending the deeper thesis that time makes a difference for modalities because the existence of powers at a time impose formal constraints on the structural conditions governing the accessibility relation between the actual and the possible. Some alternative states of affairs are not genuine possibilities in the actual world, he argues, because of the powers that obtain in the actual world. Normore, moreover, roots his defence of this thesis in the medieval debate between Scotus and Ockham over whether what is possible is possibly actual (Scotus maintained 'no it need not be', whereas Ockham maintained 'yes it had to be').

Normore structures his argument around Scotus's and Ockham's disagreement regarding Auriol's Principle, *If A is F and A can be not-F, then A can change from F to not-F.* This reduces the question to whether or not the power to F can exist despite never being actualized. Scotus, and anyone who rejects Auriol's Principle, hold that the power can so exist; Ockham, and anyone who accepts Auriol's Principle, hold that no power could so exist. To defend Scotus's position, Normore needs to pull up a couple of distinctions from the medieval context. First, he suggests that Scotus needs a distinction (developed by Ockham) between a proposition being assigned the true relative to a world *secundum vocem* and a proposition that is true because it is grounded in that world *secundum rem* (72). The basic idea is that there is a difference between a proposition whose truth is relative to the actual world and one that is made true by the actual world, and this allows us to see that, if a proposition is grounded in a world *secundum rem*, it is necessarily true in that world even though it is not generally true that, if a proposition is true in a world, it is necessarily true in that world.

This idea of grounding in a world *secundum rem* is then propped up, according to Normore, by the doctrine of causal powers (75). To see this, we need to reframe Auriol's Principle to read *If A is F and A can be not-F, then A has the power to change from F to not-F* and also recognize that Spencer's counter-examples to the 'Poss-ability Principle' give us cases where A is F and can be not-F even though it lacks any power to change from F to not-F. Normore illustrates this with an example taken from rational agency: 'to be free, an agent needs to be able to act differently from the way it does in fact act, but there is no relevant time at which it is able to act differently from the way it does act.' When this occurs, argues Normore, there obtain *instantes* of nature (indices within things that reflect the priorities in nature, as opposed to the priorities in time, between them) that relativize the alethic evaluations of propositions regarding the agent's action. There is now a prior *instans* of nature according to which the agent is able to act differently than it does and now a posterior *instans* of nature according to which it is not able to act differently. But because these *instantes* revolve around priorities of nature rather than priorities of time, they can coexist at any given time and nonetheless function as different indices against which the truth of propositions relativizes.

> If we look closely as Scotus's device of *instantes* of nature at a given time... we find that at time t, relative to a prior *instans* there is the will with its power to act or not act. At time t, relative to a posterior *instans* it acts, and so if we posit it to be not acting at that posterior *instans* of nature, we can derive the contradiction that it acts and does not. Scotus takes it as sufficient for it being possible that the will not act at t that there be some *instans* of nature at t such that if we posit it not acting relative to that *instans*, we get no contradiction. (79)

In the context of causal powers, rather than the free choices of rational agents, the causal powers themselves are the *instantes* of nature propping up the falsity of the consequent of Auriol's Principle. Thus, the contradictory modal statements are thereby banished as irrelevant to making the assessment of what is really possible at a given time in a given world because the powers within objects provide a robust metaphysical principle restricting possibilities to truths relativized to those powers. In this way, Normore gets a link between modality and time akin to this provided by the Principle of Sufficient Reason or the Principle of the Necessity of Origins but with roots that contemporary philosophers of science should find more plausible. But whether or not the friends of powers are able or willing to take on Scotus's doctrine of *instantes* of nature to support their modal semantics remains an open question.

2.3. 'Aristotelian Powers, Mechanism, and Final Causes in the Late Middle Ages', by Henrik Lagerlund

Henrik Lagerlund explores the topic of final causality in the High and later Middle Ages. He argues that the seventeenth-century mechanists weren't the only ones critiquing and rejecting final causality. There were earlier figures who developed a form of mechanical materialism that eschewed final causes, most notably William of Ockham and John Buridan (86).

Lagerlund begins with the way that Ockham and Buridan in the fourteenth century understood the mereology of the body. Bodily substances were composed of essential parts and integral parts. Essential parts were its metaphysical constituents, its matter and substantial form. Integral parts were its various extended bits. This distinction generated a metaphysical divide between material objects with extended substantial forms and simple, immaterial substances like God, angels, and the human soul. And this divide raises a number of philosophical puzzles for the entities on either side of it. Of special concern to Lagerlund is the numeric identity and unity of material substances across time. Lagerlund shows how Buridan in particular struggled to make sense of the identity and unity of material substances through time. In the end, Buridan could only say that material substances are successively identical through time; they are not totally or partially identical (87).

Final causality, according to Lagerlund, had no place within this medieval mechanical materialism because Ockham and Buridan strove to reduce their ontological commitments. This allowed them to restrict powers and final causality to the mind alone, much as Descartes would a couple of centuries later (88–89). Ockham, and later Buridan, were essentially worried about two things, according to Lagerlund. First, the coherence of the temporal priority of the final cause, qua cause, over the effect and the identification of the final cause with the not-yet-existing effect: 'Buridan argues that every cause is naturally prior to the caused thing, but the end is not naturally prior. The problem is thus how something that is posterior to its effect can be a cause' (89). Second, the priority of God over nature: 'the cause of God's action of creation is the world being created. But this implies that things inferior to

God are causes of His actions, which is clearly absurd. Buridan takes this to be a knock-down argument of final causality' (89).

But even though Ockham and Buridan had no room for final causality, Buridan still felt the need to hold onto powers in the weak sense of dispositions. According to Buridan, dispositions were special kinds of qualities that underlay an object's natural capacities (90–91). Nevertheless, maintains Lagerlund, this does not smuggle causal powers in any metaphysically objectionable sense back into Buridan's medieval mechanism because it is possible to read Buridan as meaning by 'dispositio' nothing more than the mere arrangement of its material, integral parts (91). So, even though dispositions make an appearance among the medieval mechanists, it does not make them Aristotelian or compromise in any way their anti-Aristotelian and mechanistic positions.

2.4. 'Agency, Force, and Inertia in Descartes and Hobbes', by Deborah Brown

Deborah Brown is looking at how Hobbes and Descartes used the language of 'tendency' in their natural philosophies. She contrasts the problems that arise for Hobbes as he tries to reduce tendencies away with a Descartes, for whom 'tendency talk is not a mere *façon de parler*' but rather is 'real and causally explanatory' (96), and who strives to incorporate inherent tendencies into his broader mechanistic ontological framework. The resulting interpretation of Descartes sees him as much closer in his conception of natural laws to Nancy Cartwright rather than to David Lewis. One of the real benefits of this interpretation, claims Brown, is that it 'might just help to demystify Descartes's references to active forces' (97).

Brown begins with Louis de La Forge and his slide from Descartes's law of inertia to his occasionalist claim that motive force is not a constituent of the moved body on the basis that motion is only a mode according to Cartesianism. This overlooks Descartes's commitment to active forces (*conatus*) in nature, a mistake encouraged by Descartes's comments regarding his third law of motion and the belief that Cartesian bodies are inert, Brown concedes, but a mistake nonetheless. The root of the mistake is the ease with which readers slip from the idea that bodies are *inertial* to the idea that bodies are *inert* (102).

With that mistake diagnosed and interpreters' presumption about Descartes's lack of commitment to active forces softened up, Brown then turns to the argument for the interpretation that *conatus* are tendencies and the laws regarding their unfolding are tendency laws. After establishing Descartes's commitment to tendencies in his three laws of motion, she argues that we have no real basis for not interpreting them as real, inherent properties of bodies because there is nothing in Descartes that in fact suggests otherwise, and because tendencies play a causal role within Descartes physics even when they are 'unfulfilled' (104). Brown makes a strong textual case for her claim that tendencies play a causal role in Descartes physics by carefully analysing the claims Descartes made regarding tendencies toward motion and showing that typical reductivist readings of them cannot do justice to the conclusions Descartes draws from them (104–9). Furthermore, Brown argues, this commitment to the reality of tendencies is compatible with Descartes's well-known rejection of final cause playing any role in natural philosophy.

Having thus established that Descartes was a realist about tendencies who nevertheless remained committed to a non-teleological, mechanistic account of nature, Brown contrasts this Cartesian picture with Hobbes's reductive mechanics that eliminated forces and tendencies by equating them with actual motions, including even the 'force of a body at rest' (111). On Brown's analysis of Hobbes, 'Hobbesian tendencies are real, causal powers that are neither dispositions nor potentialities' which 'belong to the category of motion' (112). The drawback, however, is that Hobbes cannot maintain a categorical difference between bodies at rest and bodies in motion, between agents and patients in other words, which entails that Hobbes cannot properly analyse cases in which equipotent forces offset one another, 'stalemate cases' in Brown terminology. The Hobbesian reductive analysis cannot work here because it requires the patient, who is being equally acted upon by opposing agents in the stalemate case, to be *per impossibile* simultaneously moving in both directions. Brown then turns to consider how Descartes's adoption of real tendencies can be applied to the Cartesian mind and why it avoids the pitfalls awaiting Hobbes's reductivist account.

2.5. 'The Ontological Status of Causal Powers: Substances, Modes, and Humeanism', by Benjamin Hill

Benjamin Hill is seeking to initiate deeper contemporary discussion of the ontological challenges that drove early modern philosophers (namely, several early Cartesians, Berkeley, and Hume) to accept the negative thesis of occasionalism, that no physical object can truly be an efficient cause. He argues that we should be looking past Hume and his empiricist's approach to secondary causation to bring the core metaphysical issues he believes are still lingering into sharper focus.

Hill walks us backwards from Hume's empirical critiques of powers in the *Enquiry* and *Treatise* to Locke's presentation of the 'popular' view that experience leads us to postulate powers as a response to occasionalism. This, he suggests, reveals the early modern debate about causal powers to have tracked not the divide between scholastics and mechanical philosophers but the divide between realists and occasionalists and revolved around a confusion between them regarding what was an underlying question of the debate. For the occasionalists, it was not really about whether or not causal powers did exist but about explaining how they could exist (121). This leads Hill to explore the metaphysical worries animating seventeenth-century occasionalists.

Beneath the four central arguments presented in favour of occasionalism, Hill identifies three metaphysical worries that, he says, ought to be of concern to contemporary practitioners: the conceptual clarity regarding the category of causal powers; the vaguenesses surrounding characterizations of and boundary conditions for an effect (Hill calls it the problem of full determination); and the nature and grounds of natural necessity as an ontological category. Hill then considers how the occasionalists developed these worries and why they found them so pressing within the modern, mechanistic scientific framework. He begins with the lack of conceptual clarity and argues that the basis for Malebranche's rejection of the idea of power was very different from (and much more powerful than) Hume's empiricist's attack precisely because it focused on the inadequacies internal to the concept of a causal power (134–35). Then he turns to showing how Malebranche's No Knowledge Argument and La Forge's No Transfer Argument for occasionalism exploit the worries about fully and properly specifying the extent and boundaries of the operations of a causal power. Conceptually speaking, every effect must be made fully determinate when it is actualized. This requires wading through a very complicated set of intersections, and Malebranche was only able to comprehend this as a set of choices and decisions by a knowing agent, hence his No Knowledge Argument. But the idea of agency limiting Malebranche's conception is a red herring, according to Hill, because the real pressure on the realist comes from the need to make sense of the complexity of the many intersections by making effects (and powers) properties that emerge from states of affairs involving many substances and relationships rather than being inherent or per se properties of substances (138–40). To avoid this 'by connecting the full determination of a power's activation to the force or energy necessary to activate the power' (140) is, Hill argues, to run right into La Forge's No Transfer Argument. This argument pressures the realists by forcing them to conceptualize force or energy into the category of a 'special' substance, which threatens, among other things, a vicious infinite regress, or into the category of mode, property, or state of a substance, which undercuts it, fulfilling the causal role realists require of it (143). The final worry explored by Hill was that involving natural necessity. He argued that this reached back to the medieval debates between voluntarists and intellectualists regarding the limitations on divine omnipotence. Their debate, Hill suggests, came down to whether or not natural necessity needed a metaphysical ground. Malebranche, following the voluntarists, emphasized that the only thing conceivably capable of grounding natural necessity is the divine will rather than the Divine Ideas. But the core problem is even worse for contemporary realists than it was for the medieval and early modern intellectualists in that they had God to fall back on but we don't, and trying to identify a properly naturalistic grounds for systematically supporting natural necessity seems much harder to get off the ground.

2.6. 'The Case against Powers', by Walter Ott

Walter Ott presents much the same take-home message as Benjamin Hill— that there is a central, underlying metaphysical worry that animated early modern thinking against causal powers that is still unresolved. But he emphasizes a different, though related, worry—the spectre of power holism embedded in the Aristotelian's commitment to intrinsicality. Ott sees this as the worry expressed in Descartes's 'little souls' argument against powers, which he sees as an important precursor to Neil Williams's recent 'problem of fit' objection to neo-Aristotelian causal powers (153). And it is this 'problem of fit' that highlights the connection between intrinsicality and powers holism.

Williams's 'problem of fit' objection is squarely aimed at dispositional essentialism: assume that powers are intrinsic properties directed at their manifestations, yet not identical with or reducible to them, and that they are essentially dispositional in this way; recognize that powers come in 'bundles' (155) that are often distributed among several different substances (for example, fire has the power to burn paper and paper has the power to be burned by paper, at least within an oxygen-rich environment); and then we can ask the simple question, 'why are these powers bundled together and not others?' Dispositional essentialists need to provide an answer here, but it is not clear what theirs would look like or how they could answer it. For any bundle of powers, it is 'perfectly mysterious', miraculous even, says Ott (156), why they bundled together, and are so perfectly suited to one another, and why they cannot be bundled otherwise. But the mystery is even deeper, and the problem worse, when we recognize that background conditions and a myriad of laws of nature (themselves intimately connected to many other bundles of properties) directly impact the nature of powers: virtually every power is bundled with every other power within a possible world. This leads, Ott argues, directly to the absurdity of 'power holism' (160–61). Against a rich version of dispositional essentialism that requires integration of causal powers one with another, like Hilary Kornblith's (1993), Ott's worry seems especially well placed. 'If power holism is the cure,' Ott colourfully states, 'one might be forgiven wanting the disease back' (161).

According to Ott, however, the lesson to take from the 'problem of fit' is not that Humeanism is the proper position. He suggests instead that we return to a mitigate powers ontology like that found in Boyle and Locke. The idea here is that powers are really relational properties that cannot be reduced to any one of their relata (155); they simply 'are internal relations among the intrinsic properties of things' (165).

2.7. 'The Return of Causal Powers?' by Andreas Hüttemann

Andreas Hüttemann disagrees with Hill and Ott regarding the relevance of the early modern critiques of causal powers for contemporary practitioners. He argues that the contemporary acceptance of powers and dispositions is insulated against the early modern criticism because the emergence of powers nowadays is not a 'revival of' or 'return to' the Aristotelian or scholastic version of causal powers. Hüttemann traverses two lines of argumentation in his defence of the contemporary metaphysics of powers.

First, he maintains that the early modern critics utilized a version of causation that because it was rooted in the doctrine of substantial forms was quite strong and restrictive and that, consequently, their criticisms don't apply to contemporary notions of powers, which utilize a counterfactual conception of causation. He suggests further that the early modern criticisms are in fact compatible with this counterfactual conception of causation, which someone like La Forge accepted as 'quasi-causation' (172).

Then, Hüttemann turns in a different direction to defend Nancy Cartwright and Jeremy Hardie's use of the Extrapolation Argument in favour of the postulation of dispositions and powers. Cartwright and Hardie argue that the inferences involving extrapolation from one case to another, especially when the cases involve important dissimilarities, are only possible because inherited tendencies and powers ground and 'fix' a capacity between the different scenarios. According to Hüttemann, what is happening in these situation is that there are 'different factors or tendencies that contribute to the actual behaviour that we want to explain' (180), and so long as there are 'laws of composition that allow us to estimate the contributions of the various factors' (181), we can ground such extrapolative inferences as Cartwright and Hardie want. This allows for a distinction between the complete manifestation of a power and the partial manifestation of a power such that in cases where a power is only partially manifested it is nevertheless a full and proper instantiation of the power equivalent to its instantiation in cases where it is completely manifested (180–81). In order for this to make sense, Hüttemann says, it is not necessary for the tendencies, powers, and dispositions undergirding an extrapolation to be causal. Indeed, he says, such dispositions are often 'infinitely multitrack', which removes them still further from the classical notion of causal. Thus, contemporary references to dispositions and powers should not be too quickly subsumed under the early modern philosophers' conception of causal powers (182).

2.8. 'Qualities, Powers, and Bare Powers in Locke', by Lisa Downing

Lisa Downing focuses on the important issue of the metaphysics of Locke's primary–secondary qualities distinction. In recent years this has returned as a topic of scholarly contention. Downing is concerned by the anti-realist trends in recent work on the metaphysics of Locke's primary–secondary qualities distinction (186), and she is keen to defend the claims that Locke was 'putting forward a kind of *realism* about secondary qualities' and that his realism does not readily appear to be a reductive form of realism (195).

Downing begins with the traditional claim that Locke's distinction was driven by his understanding of matter theory within the new science, like many others in the seventeenth century (189). From this perspective, she criticizes recent work on the nature and priority of primary qualities, which fail to root the primary in a metaphysical base (Stuart 2013) or connect them to the metaphysical base in the wrong way (Jacovides 2017; Pasnau 2011). Next, she turns toward explaining her own understanding of the subordinate status of the secondary qualities, which brings Downing to Locke's claim that

secondary qualities are 'mere powers' (194) and what this meant metaphysically to him. It is here that Downing begins her defence of Locke's non-reduc-
tive realism about secondary qualities.

Her approach toward defending the realist position starts by critiquing the
anti-realist and idealist interpretations recently pouring forth. She quickly
knocks down Jacovides's (2017) interpretation that Locke is only offering a
semantics for secondary qualities terms (196), but then turns to attack the
relational account of Ott (2009). Ultimately Downing cites 'the unattractiveness of the view that relations simply do not exist, plus the deep connections
Locke sees between modes and relations' as 'sufficient grounds' for discount-
ing the robustly anti-realist text in the *Essay* (II.xxv.8, 322) and for pushing
any viable interpretation of the metaphysics of secondary qualities into a
realist direction. So, she next turns to Stuart's (2013) notion of secondary
qualities as degenerate powers (198). Downing identifies a host of problems
with this notion, however, including the consequences: that secondary qualities are actualities, not potentialities (199); that secondary qualities are not
stable and persistent, which also undermines the stability and persistence of
kind membership (199); and that this move to degenerate powers does not
really avoid the problem of making them out to be relations, which don't
really exist (199–200).

After securing that the anti-realist interpretation is unacceptable, Downing
considers the case for attributing reductive realism to Locke (Pasnau 2011;
Ott 2009). She does not find the reasons cited for requiring us to attribute
reductivism to Locke persuasive. But her biggest concern seems to be with
the absence of details regarding how the reductions are supposed to actually
take place. For Downing the central notion behind Locke's position is that
the secondary qualities that supervene on bodies as mere powers are not *res*
in the technical scholastic sense of being separable existents (202). But the
further claim that secondary qualities are not existents in any other sense is
not obviously excluded by Locke, so why should we attribute anything
stronger to him seems to be what Downing is saying. But there is more to be
said for the realist, anti-reductivist interpretation, according to Downing,
when we examine what is necessary to prop up the epistemological claims
Locke maintains regarding secondary qualities. The adequacy of our simple
ideas requires that the secondary qualities be real existents: 'For the claim
that all our simple ideas are adequate to come out true, it is necessary that
our ideas be aimed at powers, that there *be* powers, and that those powers be
completely known' (203).

2.9. 'Hume on Causation and Causal Powers', by Peter Millican

Peter Millican returns us to the issue of how best to interpret Hume's iconic passages on causation and causal powers and aims to cut through the various interpretations by fixing twelve 'key points' (208) and arguing that a reductivist reading makes best sense of them. His twelve key points are:

1. Whether A causes B is an objective matter of fact, and causes can be discovered by systematic investigation (209).
2. Causes are understood to be prior and contiguous to their effects (209).
3. The principle component of the concept of causation is necessary connection, which is essential to it (210).
4. Causal necessity is not the same as conceptual necessity (211).
5. Hume is a convinced determinist, although his basis for this is unclear (212).
6. Necessary connection is one of a family of 'power' terms, which Hume treats as virtually synonymous in this context (213).
7. Understanding 'power' terms involves having a certain simple idea, which is copied from a corresponding impression of reflection (214).
8. The corresponding impression of reflection arises from observed constant conjunction and the consequent tendency to draw inductive inferences (215).
9. Hume provides two definitions of cause (216).
10. Hume also provides two definitions of necessity, which he applies to issues of 'liberty and necessity' (218).
11. When the two definitions of cause come apart, constant conjunction dominates (219).
12. In *EHU*, Hume recognizes more sophisticated causal relations than in *THN*, mediated by quantitative powers and forces (221).

With these twelve points regarding Hume's theory fixed, Millican turns toward adjudicating between reductivist, subjectivist, and projectivist interpretations. First, Millican attacks subjectivist interpretations on the grounds that they emphasize melodramatic passages in tension with Hume's more considered claims, especially the first definition of necessity (key point no. 10). Indeed, Hume himself, Millican argues, backed away from the subjectivist ideas by the time he came to write the *Enquiry*: 'His omission of the stridently subjectivist declarations of the *Treatise*, therefore, was both deliberate and appropriate' (227). Millican backs up the critical comments about subjectivism with 'a plausibly Humean account of what his "impression of power or necessary connection" might be' (227). Then he turns to projectivist interpretations. Here, he argues that projectivist readings can be accommodated by the reductivist reading he is defending (230–32). After that, he turns to the 'New Hume', who allegedly accepted 'thick' causal powers, which push beyond the two definitions of cause. But Millican emphasizes that Hume did accept causal powers in some thinner sense, powers that reduce to causal structures in the world that allow the discovery of laws and enable predictive success.

2.10. 'Resurgent Powers and the Failure of Conceptual Analysis', by Jennifer McKitrick

Jennifer McKitrick returns us to contemporary times and examines the causes moving many philosophers to pull Aristotelian powers out of history's dustbin—the failure to reduce or eliminate dispositional ascriptions from philosophical and scientific discourses. Although many see this failure as grounds for rejecting Humeanism and return to Aristotelianism, McKitrick argues that only a more moderate reaction is warranted. She argues that restricting analysis to fundamental dispositions and adding a condition requiring the power ascription be grounded in or made true by the fact that the object possesses that fundamental disposition is the better reaction.

McKitrick canvasses the main twentieth-century attempts to reduce or eliminate dispositional talk. She begins with the logical positivists' attempt to replace dispositional talk with material conditionals. Then, after briefly considering Ryle's version, she turns to Goodman and the move to replace material conditional analyses with 'stronger-than-material conditionals, such as causal implication or counterfactual conditionals backed up by natural kinds and laws of nature' (245). Next, she turns to Lewis's possible world semantics and concludes with a presentation of the 'Simple Counterfactual Analysis', which states 'x has D $=_{df}$ If x were in C, x would exhibit M' (250). The legacy of these attempts, according to McKitrick, is the failure of the any of these analyses to actually get the right result because the relationship between antecedent and consequent is never narrow enough to exclude accidental coincidences.

But things get even worse in the late 1990s and early 2000s, maintains McKitrick, with the emergence of a slew of counterexamples to the Simple

Counterfactual Analysis, mimics, masks, and finks. Mimics are objects that seem to manifest a disposition without actually possessing the disposition in question (251). These counterexamples show that the counterfactual is not sufficient for ascribing a disposition to an object. Masks and antidotes show that the counterfactual is not necessary for ascribing a disposition to an object because it prevents the disposition from masking even though it is being activated as it ought to be. Finks and other alterations, like Martin's famous electro-fink, remove the disposition at the precise moment that its manifestation would normally be activated.

These counterexamples have pushed many philosophers to explore more sophisticated versions of the counterfactual analysis capable of avoiding the finks, masks, and mimics. None of these, however, are satisfactory, according to McKitrick. Adding some kind of *ceteris paribus* constraint to the counterfactual seems to only be adequate when it packs reference to dispositions into its analysans, which undercuts the whole reductivist and eliminativist programme (256). David Lewis's attempted revision puts structural properties at the base of an object's causal powers and requires the preservation of these causal bases in the analysis. But this too threatens to undercut the whole reductivist-eliminativist model by packing into its analysans a property which has causal oomph (258). McKitrick ends her tour of reductivist-eliminativist attempts with Manley and Wasserman's Proportionality Account, 'N is disposed to M when C if and only if N would M in some suitable proportion of C–cases' (259). Although this analysis goes farther toward resolving the problems with reductive, counterfactual analyses than others, it still struggles with intrinsic fink counterexamples and with the need to refer to infinite sets of possibilia when specifying the C–cases. 'PROP relies on the Lewis-Stalnaker semantics for counterfactuals, so as suggested earlier, the price of reducing disposition ascriptions might be a multiverse of possible worlds' (261). McKitrick's tour is a tour de force of the evolution and current state of reductivist-eliminativist attempts to avoid dispositions.

Despite these problems with providing analyses of dispositional ascriptions in terms of counterfactuals, McKitrick recognizes that 'there is still an important connection between dispositions and counterfactuals. A thing's disposition is its property of having a certain kind of counterfactual hold of it' (265). But she advocates restricting counterfactual analyses to fundamental dispositions and powers and requiring that they be 'made true, or grounded by, the fact that the object has that power' (265).

2.11. 'Causal Powers and Structures', by Brian Ellis

Brian Ellis provides a detailed and systemic overview of his version of dispositional essentialism. Ellis is famous for having developed and defended a mixed ontology for scientific realism (Ellis 2001). His is a robustly Aristotelian ontology that involves a mix of categorical and essentially dispositional properties inhering as universal in individualized entities. In this contribution, Ellis briefly defends this sort of ontology by arguing that it, or something very much like it, is necessary to provide an account of the system of reality discovered through modern science. It is, moreover, entirely adequate to the job of accounting for the ontology of modern science (274). So really, what more could philosophers ask for?

He then turns to consider three objections to his ontology. The first is what other contributors will call the 'directedness problem', which is the idea that powers are directed at their manifestation in a way analogous to the directedness of intentionality. The second is what other contributors call the 'intrinsicness problem', which is the idea that causal powers are intrinsic to or inhering in their subject. The third is what other contributors call the 'necessity problem', which is the idea that there is something important and distinctive about metaphysical necessity vis-à-vis logical necessity.

Next Ellis turns to more detailed outlines of the central features of his realist ontology. He presents first his account of what constitutes a physical causal process for him—'for one system to impact upon another, there must be some physical causal process involved, and typically the impact will involve a transfer of energy of some kind from the impacting body to the body impact upon' (277)—and invokes a distinction between positively and negatively acting causes to accommodate cases of causal 'interventions' (277). Then he takes up what a 'law of action of the causal power' should look like and the role played by the categorical properties of time and location in a causal law. He further argues against the suggestion that they are 'quiddities, and therefore unreal' (280). Therefore, to accommodate spatial and temporal properties as well as structural properties that directly impact the action of causal powers, Ellis adopts categorical properties into his ontology alongside the dispositional properties of fundamental particles. Ellis is, he avers, a categorical realist as well as a dispositional realist and holds what he thinks is a 'fundamentally Lockean position' (281).

Readers coming to this volume without strong grounding in the contemporary literature regarding dispositional essentialism will especially benefit from Ellis's elegant and detailed recount of his version of the ontology. But even experts will appreciate the clarity with which Ellis summarizes the main points motivating the commitment to dispositional essentialism and the firm defence of them Ellis gives here.

2.12. 'Induction and Natural Kinds Revisited', by Howard Sankey

Howard Sankey returns to his previous work on this topic and reconsiders a special issue closely connected with causal powers—the problem of induction. Here he addresses a deep version of problem of circularity originally raised by Psillos, which afflicted his earlier work 'Induction and Natural Kinds' (1997), and argues that the circularity can be avoided. The key is recognizing certain epistemically externalist results of the Megaric consequences of the commitment to dispositional essentialism. Circularity can be avoided, Sankey argues, because it is the way the world is, rather than the inductive inference itself, that grounds the reliability of the inductive inference in his previous account.

The several sections of Sankey's paper recap for readers his previous work resolving the problem of induction and the charge of circularity lodged against it. Enumerative induction can be externally justified because it is a reliable belief formation mechanism. The charge of circularity beneath this resolution is that it uses an ampliative inference (inference to the best explanation, which justifies the metaphysical commitment to dispositional essentialism) to justify an ampliative inference (inductive inference) (284). The key to avoiding this form of circularity is to recognize the uniformity of nature, Sankey says. The commitment to nature's uniformity is a metaphysical claim whose truth grounds the reliability of our inductive inferences. We use inference to the best explanation, however, only within the context of discovery and not within the context of justification when we ground the reliability of induction, argues Sankey (296).

What are doing the work for Sankey here are the Megaric consequences of his adoption of Ellis's dispositional essentialism. The uniformity in question is one that stretches across possible worlds: 'nature is uniform in the precise sense that there are natural kinds whose members all possess a shared set of essential properties' (292). As Normore emphasized in his contribution to this volume, the significance of this commitment lies in how the possible and the temporal intersect through restrictions placed on the accessibility relation between the actual and the possible. *Ipso facto*, when considering questions about the future behaviours of objects, which is how Sankey understands the problem of induction to be, the uniformity of nature can ground the reliability of beliefs about those future behaviours precisely because the domain of possibility is restricted to those worlds accessible to the actual world, which is fixed by the commitments of dispositional essentialism. As Sankey puts it, 'Unobserved members [and actually possible members, we might add] of a kind possess the same essential properties as observed members of the kind. The fact that observed and unobserved members of a kind possess the same essential properties is what makes induction reliable' (292). Although Sankey limits himself in this paper to inductions 'underpinned by the existence of a natural kind' (294), we believe that focusing on the Megaric basis beneath Sankey's move would allow friends of powers to extend Sankey's approach to causal powers more generally, including those of individuals and instances in which natural kinds play no apparent role.

3. Questions for Moving Forward

Moving forward beyond this book, it seems important to state a few issues or questions that we have not touched upon here or questions that are generated by the articles published below. Naturally, the scope of this book is limited to the period starting with the late medieval and early modern discussions of powers and moving into contemporary philosophy of science and metaphysics. We say very little about the concept of power or causal powers before Aquinas—even though some articles touch upon Aristotle as the most obvious background to our period. There is much more to be said about late ancient discussions as well as earlier medieval discussions. We also ignore the Arabic philosophical discussions, which were essential to Aquinas as well as many others in the Latin tradition and contain very interesting discussions of Aristotelian powers as well. There is, hence, much more to be said if we look back into history.

There are a number of unresolved questions about the later Middle Ages that future studies will have to address. The fourteenth century seems to have contributed a new trend in relation to powers, which is very similar to the negative attitude we can find in the seventeenth century. This poses some new and interesting questions, such as how were causal powers construed in the *via moderna* tradition following Ockham and Buridan, what is the relation between the fourteenth-century criticism of powers and the later seventeenth-century one, and what influence did sixteenth-century discussions of final causality and the scientific method have on the subsequent seventeenth-century discussion of powers and laws of nature? Answering such questions will give us a better understanding of the seventeenth-century conception of powers and fill in our history a bit more.

And yet, not only are there many historical questions to ponder if we look forward but also there are unresolved issues about which there is a lot to be learned by blending historical and conceptual perspectives. One important issue, in our view, is the place of powers in the current scientific image of the world. While the critique of powers in the seventeenth century was associated with the emergence of a, by and large, new framework for doing natural science, namely the so-called mechanical natural philosophy, it's not quite clear whether the current resurgence of powers is linked with developments in the sciences or whether it is due mostly to philosophical reasons. Some detailed work on the status of power-based explanations of natural phenomena would be required if we were to have a clear account of the role of powers in current science. And to what extent are the arguments against powers from within mechanical philosophy relevant to current debates? Similarly, to what extent can current arguments in favour of powers cast light on the seventeenth-century debates? Besides, an important issue is the relation between powers and laws. Are either powers or laws enough to account for natural phenomena?

We could carry on. But the readers might well be anxious to delve into the various chapters of this collection and see for themselves how the blending of the historical and conceptual perspectives can be useful (perhaps, even indispensable) in thinking about the twists and turns of the philosophical engagement with powers and their place in nature and in science. The editors wish to thank Peter Momtchiloff from OUP for his encouragement, Deva Thomas for his patience and care with the manuscript, and Marilina Smyrnaki, from the University of Athens, for preparing the index of the book.

References

Armstrong, D. M. 1983. *What is a Law of Nature?* Cambridge: Cambridge University Press.

Armstrong, D. M. 1997. *A World of States of Affairs*. Cambridge: Cambridge University Press.

Bird, Alexander. 2007. *Nature's Metaphysics: Laws and Properties*. Oxford: Oxford University Press.

Carnap, Rudolf. 1936. 'Testability and Meaning'. *Philosophy of Science* 3, no. 4 (October): 419–71.

Carnap, Rudolf. 1937. 'Testability and Meaning—Continued'. *Philosophy of Science*, vol. 4, no. 1 (January): 1–40.

Cartwright, Nancy. 1983. *How the Laws of Physics Lie*. Oxford: Clarendon Press.

Cartwright, Nancy. 1989. *Nature's Capacities and Their Measurement*. Oxford: Clarendon Press.

Dretske, Fred I. 1977. 'Laws of Nature'. *Philosophy of Science* 44, no. 2 (June): 248–68.

Ellis, Brian. 2001. *Scientific Essentialism*. Cambridge: Cambridge University Press.

Goodman, Nelson. 1983. *Fact, Fiction, and Forecast*, 4th ed. Cambridge: Harvard University Press.

Harré, Rom. 1970. 'Powers'. *British Journal for the Philosophy of Science* 21, no. 1 (February): 81–101.

Harré, Rom, and E. H. Madden. 1975. *Causal Powers: A Theory of Natural Necessity*. Oxford: Basil Blackwell.

Jacovides, Michael. 2017. *Locke's Image of the World*. Oxford: Oxford University Press.

King, Peter. 2001. 'John Buridan's Solution to the Problem of Universals'. In *The Metaphysics and Natural Philosophy of John Buridan*, ed. J. M. M. H. Thijssen and Jack Zupko, 1–28. Leiden: Brill.

Kornblith, Hilary. 1993. *Inductive Inference and its Natural Ground*. Cambridge: MIT Press.

Leibniz, Gottfried W. 1989. *Philosophical Essays*. Ed. and trans. Roger Ariew and Dan Garber. Indianapolis: Hackett Publishing.

Lewis, David. 1973. *Counterfactuals*. Cambridge: Harvard University Press.

Lewis, David. 1997. 'Finkish Dispositions'. *Philosophical Quarterly* 47, no. 187 (April): 143–58.

Locke, John. 1990. *Drafts for the Essay Concerning Human Understanding and Other Philosophical Writings*. Ed. Peter H. Nidditch and G. A. J. Rogers. Oxford: Clarendon Press.

Mackie J. L. 1977. 'Dispositions, Grounds and Causes'. *Synthese* 34, no. 4 (April): 361–70.

Martin, C. B. 1994. 'Dispositions and Conditionals'. *Philosophical Quarterly*, 44, no. 174 (January): 1–8.

Mill, J. S. 1911 (1843). *A System of Logic: Ratiocinative and Inductive*. 8th ed. London: Longmans, Green & Co.

Molnar, George. 2003. *Powers: A Study in Metaphysics*. Oxford: Oxford University Press.

Mumford, Stephen. 2004. *Laws in Nature*. New York: Routledge.

Mumford, Stephen, and Rani Lill Anjum. 2011. *Getting Causes from Powers*. Oxford: Oxford University Press.

Newton, Isaac. 1999. *The Principia: Mathematical Principles of Natural Philosophy*. Trans. I. Bernard Cohen and Anne Whitman, assisted by Julia Budenz. Berkeley and Los Angeles: University of California Press.

Newton, Isaac. 2004. *Philosophical Writings*. Ed. Andrew Janiak. Cambridge: Cambridge University Press.

Ott, Walter. 2009. *Causation and Laws of Nature in Early Modern Philosophy*. Oxford: Oxford University Press.

Pasnau, Robert. 2011. *Metaphysical Themes 1274-1671*. Oxford: Oxford University Press.

Plato. 1997. *The Complete Works*. Ed. John Cooper. Indianapolis: Hackett Publishing.

Prior, Elizabeth, Robert Pargeter, and Frank Jackson. 1982. 'Three Theses about Dispositions'. *American Philosophical Quarterly* 19, no. 3 (July): 251-7.

Quine, W. V. 1974. *The Roots of Reference*. The Paul Carus Lectures, 14. LaSalle, IL: Open Court.

Ramsey, Frank P. 1978. 'Universals of Law and of Fact'. In *Foundations: Essays in Philosophy, Logic, Mathematics and Economics*, ed. D. H. Mellor, 128-32. London: Routledge & Kegan Paul.

Reid, Thomas. 2011 (1788). *Essays on the Active Powers of Man*. Ed. John Bell. Cambridge: Cambridge University Press.

Shoemaker, Sydney. 1980. 'Causality and Properties'. In *Time and Cause: Essays Presented to Richard Taylor*, ed. Peter van Inwagen, 109-35. Dordrecht: Reidel.

Stuart, Matthew. 2013. *Locke's Metaphysics*. Oxford: Oxford University Press.

Tooley, Michael. 1977. 'The Nature of Laws'. *Canadian Journal of Philosophy* 7 (4): 667-98.

1

The Inherence and Directedness of Powers

Stathis Psillos

Powers are supposed to be intrinsic properties of their bearers and to be directed towards their manifestations. But the two central features of powers are in some tension with each other. The possession of a power by a bearer *intrinsically* implies nothing outside the bearer and the power it possesses. But the *directedness* of the power implies something outside the bearer and the power it possesses, viz. that towards which the power points. How can it be that a power has its being in the substance that bears it and at the same time be towards something else? In this chapter I aim to offer a systematic account of the Aristotelian–Thomist way to deal with this problem and to highlight how this account might cast light on how the problem has to be dealt with currently. I will argue that the Aristotelian–Thomist way to tackle the inherence and directedness of powers constitutes a coherent, if metaphysically inflated, way to bring them together. It will transpire, however, that powers have to be seen as possessing a dual character: they are both qualities and relatives. This duality explains why they inhere in substances (qua qualities) and have their being towards something else (qua relatives).

In Section 1, I will outline the current conceptions of powers as being intrinsic and directed. In Section 2, I will offer an account of the marks of Aristotelian powers. Section 3 will discuss Aristotle's distinction between powers-in-potency and powers-in-act. In Section 4, I will present the view that powers imply relatedness. Section 5 will move on to the Thomist account of the towardness of powers. Section 6 will defend the view that powers have a dual nature. Section 7 is going to claim that, though powers are real, the distinction between their qualitative nature and their towardness is a conceptual one. Finally, in Section 8, we will take stock and see the bearing of all this to modern conceptions of powers.

Stathis Psillos, *The Inherence and Directedness of Powers* In: *Reconsidering Causal Powers: Historical and Conceptual Perspectives*. Edited by: Benjamin Hill, Henrik Lagerlund, and Stathis Psillos, Oxford University Press (2021).
© Stathis Psillos.
DOI: 10.1093/oso/9780198869528.003.0002

1. Powers: Directedness versus Inherence

The manifestation of a power is what results from the exercise of the power. Characteristically, powers are possessed by their bearers even when they are not exercised. Powers, then, are distinct from their manifestations. Yet, they are taken to be *directed* towards them. But then they are directed to something 'beyond themselves'.

The late George Molnar (2003) took directedness to be akin to intentionality. He called it 'physical intentionality' and claimed that it separates powers from non-powers. As he put it, 'A power has directionality, in the sense that it must be a power for, or to, some outcome. It is this directedness that provides the prima facie distinction between powers (dispositions) and non-powers' (57). In fact, for Molnar directedness is 'constitutive' of a power: 'A power's type identity is given by its definitive manifestation' (60). This definitive manifestation is 'the intentional object of a physical power'. This directed-at intentional object 'reveals the nature (identity) of the power'. 'Consequently,' Molnar says, 'the nexus between the power and its manifestation is non-contingent. A physical power is essentially an executable property' (63).

Yet, it's not quite clear what this directedness, which powers possess and non-powers lack, is. Molnar treats it as an undefined primitive: 'Directedness is not defined. My proposal is to treat physical intentionality as an undefined primitive of the theory of properties' (81).

Be that as it may, the idea that powers are directed to something 'beyond themselves' is widespread among the friends of powers. The late E. J. Lowe (2016, 107) took that as clear: 'Causal powers are, in a way, rather like intentional states, such as loving. They are "directed" at other objects of various kinds, but don't require the *existence* of those objects. But the *manifestations* of causal powers appear equally to be monadic properties of the objects in question.' Mumford and Anjum (2011, 24) agree that a power has 'a direction— that towards which it is disposed—such as fragility being a disposition towards breaking'. They go as far as to represent powers as vectors and hence to capture the directedness of a power by the direction of a vector.[1]

That's half of the story about powers. The other half is that powers are widely taken to be intrinsic properties of their bearers. Molnar (2003, 9) notes that 'powers are intrinsic properties of their bearers, so having a power is independent of the existence of any other object.' John Heil (2005, 344)

---

[1] For criticism of the vectorial representation of powers, see Pechlivanidi and Psillos 2017.

stresses that a defining characteristic of powers is that they 'are intrinsic properties of objects possessing them'. He adds: 'Dispositions are neither relations nor "relational properties." A disposition is not a relation to actual or possible manifestations or manifestation partners.' Elaborating on this view, he notes: 'Objects possess dispositions by virtue of possessing particular intrinsic properties. The nature of these properties ensures that they will yield manifestations of particular sorts with reciprocal disposition partners of particular sorts. In this regard, the dispositions "point towards" non-actual, merely possible manifestations with non-actual, merely possible disposition partners. The "pointing" is grounded in the disposition, however, not in a relation the disposition bears to anything else' (Heil 2003, 221).

The key idea behind 'intrinsicness' is that a property is intrinsic to its bearer if (and only if) its possession by the bearer does not in any way require, or depend on, the existence of any other individual.[2] Molnar (2003, 102) summed this up by saying that 'P is intrinsic to x iff x's having P, and x's lacking P, are independent of the existence, and the non-existence, of any contingent object wholly distinct from x.' Hence, the intrinsicness of powers is contrasted to their being relational properties of their bearers.

For a power P, then, to be an intrinsic property of its bearer X, it should be such that the power would be possessed by X even if there were no object other than X. But power P, qua power, is directed to something outside its bearer. The power P is always a power for something: that for which it is a power for (be it possible or actual). The problem, as I see it, is that the two constitutive features of powers—intrinsicness and directedness—seem to be in some kind of tension. The possession of power by a bearer *intrinsically* implies nothing outside the bearer and the power it possesses. But the *directedness* of the power implies something outside the bearer and the power it possesses, viz. that towards which the power points.

The directedness of powers implies or suggests some kind of relational account of powers. No matter how exactly we analyse *directedness* (and even if we don't, à la Molnar), taking it seriously requires that the power is pointing to something outside the bearer of the power. But if A points to B, A is related to B: there is no pointing without something being pointed at. Hence, directedness implies some kind of *relatedness*, though the intrinsicness of a property does not. How can a power be both intrinsic to its bearer and 'have its being' toward something outside it?

---

[2] For discussion of the various views of intrinsicness, see Yates 2016.

This, of course, is not a problem that has not been noticed. Conditional analyses of powers, for instance, take it that powers should be understood relationally, but they abandon the thought that powers are intrinsic to their bearers. Solubility, for instance, is taken to be a property of X insofar as there are solvents: for X to be soluble (in Y) it should be the case that if X is (or were to be) placed in Y (which is a solvent), X dissolves (or would dissolve). As Heil (2003, 81) noted, 'The inspiration for a relational conception of dispositions [powers] arises from...our practice of identifying dispositions conditionally, identifying them by reference to their possible manifestations.' But as Molnar (2003, 112) observed, there is a conflict between the intuition that powers are intrinsic to their bearers and the philosophical practice to analyse powers relationally.

Both Molnar and Heil (and most friends of powers) have objected to a relational analysis (let alone a reduction) of powers on the basis of conditionals. Heil (2003, 83) admits that though powers 'can be conditionally characterized in a way that invokes their actual or possible manifestations,...this does not turn dispositions into relations'. For him, 'The existence of a disposition does not in any way depend on the disposition's standing in a relation to its actual or possible manifestations or to whatever would elicit those manifestations.' Molnar (2003, 19) is equally firm: 'All truths about the powers of objects have only intrinsic properties as truthmakers.' More recently, David Yates (2016, 139) noted that, though 'powers are directed towards their manifestations', they 'do not ontologically depend upon them, and so are intrinsic to their bearers'. Yet the tension noted above does not evaporate. Directedness invites thinking of powers in relational terms while intrinsicness does not. Yates put the issue thus: 'if powers have relational essences, how can they be intrinsic?' (143).

My aim in this chapter is modest. I want to show that the tension between inherence and directedness is an issue that has been discussed by the Aristotelian friends of powers in the past and, on the basis of this discussion, to provide a *metaphysically coherent* picture of how powers can be in a subject and directed to something else. The key issue is this: powers qua qualities are inherent to their bearers; they have their being *in* them; but powers are relative too (and hence they imply *relatedness*), since their possession implies the (potential or actual) existence of those items towards which the power is directed. Powers are *both* qualities and relatives; hence they are neither pure qualities, nor pure relatives. According to the Aristotelian–Thomist view that I will explore, powers have a dual character, which explains why they inhere in substances (qua qualities) and have their being towards something else (qua relatives).

In the sequel I will follow the Aristotelian–Thomist tradition and talk of inherence (*inesse*), viz. the being of an accident (a non-substance) in a subject (a substance). As Aristotle explains in the very beginning of the *Categories*, something is said of a subject in two ways: it is either *said of* a subject but is not in it; or it is *in* a subject but is not said of it (A 1a20–1b9, 1:3). For instance, Aristotle says, 'man is said of a subject, the individual man, but is not in any subject', whereas 'the individual white is in a subject, the body (for all colour is in a body), but is not said of any subject'. He clarifies that by 'in a subject' he means 'what is in something, not as a part, and cannot exist separately from what it is in'. Inherence, then, should be taken to stand for 'being in' a subject. Now, substantial universals (species and genera), that is secondary substances as Aristotle put it, are *said of* primary substances (individuals) but are *not* in them. By being predicable of a primary substance, they make it be of a kind (for example, man, animal, etc.), thereby making it, as Terrence Irwin (1988, 58) put it, 'fit for further characterization by its inherent properties'. The way substantial universals characterize primary substances is called by Irwin 'strong predication' and is opposed to inherence, which is the way non-substantial universals (or particular instances of them, aka tropes) are *in* the primary substance.[3]

Inherent properties then are the accidents that characterize a substance of a certain kind. As Irwin (81–2) noted, 'If a universal is inherent in a particular substance, then clearly it depends on that substance, since inherence implies existential dependence.' Now, among the inherent properties of a substance some are essential for this substance, qua member of species. For instance, the power for knowledge is a necessary inherent property of humans. Those inherent properties of a substance that are essential to it are intrinsic to the substance. They are in the substance 'in virtue of its own nature [*kath' hauto*]' (A 1003a22–3, 2:1584). As Irwin (1988, 119) explained it, 'An intrinsic property must be either an essential property or derived from the essence.'

Since intrinsic properties is a subclass of inherent properties, a necessary condition for a property being intrinsic is that it is *in* a subject. The foregoing tension arises precisely because powers are inherent and directed.

---

[3] For a classic account of inherence in Aristotle, see Owen 1965. For medieval conceptions of inherence, see Pasnau 2011, 200–20.

2. The Marks of Aristotelian Powers

Aristotelian powers[4] constitute the locus and cause of activity in nature. They are posited to explain action/change ('movement and becoming'). As Aristotle (A 1048a 28–9, 2:1655; translation amended) put it, powerful [δυνατὸν] is 'that whose nature is to move something else, or to be moved by something else, either without qualification or in some particular way'.

In *Metaphysics* Book 5, Aristotle defined power (*dunamis*) as 'a source of movement or change, which is in another thing or in the same thing *qua* other' (1019a15–16, 2:1609). Hence the ascription of power requires at least a *conceptual* distinction between two things, one that acts and another that is acted upon. To illustrate this he offers the following examples: the builder has the power to build a house; in this case, the power to build is *in* the builder (who builds) and not in the house (which is being built); but the doctor may have the power to heal himself, and in this case, the power is *in* 'the same *qua* other', since the doctor (who has the power to heal) heals himself not qua doctor but qua patient, even if the doctor and the patient are numerically the same.

In a similar fashion, in *Metaphysics* Book 9, Aristotle (1046a10–12 2:1651; translation amended) defined the *primary* sense of power as 'a principle of change in something else or in itself *qua* something else'. He continued, adding that the power to be acted upon (that is, the *passive* power, or *capacity*) has its origin 'in what is itself affected of being changed' but requires being 'acted on by something else or by itself *qua* something else'. Hence, powers are active and passive, and both types are required for an account of action and interaction. In fact, the passive power (capacity) of a substance is inherent in the substance which is affected, *even though* its activation requires the active power of another substance. The same holds for active powers too. That is, the active power of a substance (for example, of the fire to heat or of the builder to build) is inherent to this very substance (for example, the fire or the builder).

So powers inhere in substances and come in pairs of active and passive. They ground and explain action in the following way: substance X *acts on substance* Y because (by nature) X has the active power to Φ and Y has the passive power to be Φ-ed. Causation is then seen (at least in its first of the four modes) as production: X *brings about a change in* Y because of its active power to Φ and because of the passive power of Y to be Φ-ed.

---

[4] The generic term Aristotle uses is *dunamis*. I will use the words 'power' and 'potency' interchangeably as translations of *dunamis*, reserving the term 'capacity' for passive power.

Aristotelian powers act by contact. When a power acts (at some time and in the required way) and if there is 'contact' with the bearer of the correlative passive power, the effect necessarily (that is, inevitably) follows (A 324b8, 1:530).

Substances possess powers even when these powers are not exercised (acting). This is a key feature of Aristotelian powers: they can exist unactualized. Hence, they are irreducible to their actualization. Aristotle takes pains to block the view that powers exist if and only if they act—this is the view he attributes to the school of Megarians (1046b29–32, 2:1653). Hence, he is keen to avoid *actualism*, viz. the view that 'a thing can act only when it is acting and when it is not acting it cannot act'. For Aristotle that X has the power to Φ and yet X is not Φ-ing is a possible state of affairs and captures *what it is* for X to have a *power* to Φ. What it is for X to possess Φ, even if Φ is not exercised is the same as not being *impossible* for X to actually Φ, *if and when the opportunity arises* (1047a28–30, 2:1653).

3. Powers in Potency versus Powers in Act

Given Aristotle's distinction between the possession of a power by a substance and its activation/exercise, can it be the case that a power of a substance can be possessed by it even though there are no actual correlative (passive) powers? The intuition here is this. The power of X to Φ (for example, of the fire to heat water) can be possessed by X even if there is no *actual* substance with the passive power to be Φ-ed (for example, even if there is no water to be heated). But the power of X to Φ cannot become active *unless* there is a Y with the power to be Φ-ed. Hence, when the power of X to Φ is exercised, there has to be a Y with the passive power to be Φ-ed.

A main task of *Metaphysics* Book 9 is to explain the sense in which a power can be possessed by a substance even though it is unexercised. In chapter 6, Aristotle (1048a29–30, 2:1655) famously noted that, though the *primary sense* of the potent [δυνατὸν] is that whose nature is to move something else, or to be moved by something else, either without qualification or in some particular way, we also speak of the potent differently and he aimed to show how and why this is so. To do this, he contrasted the potent with the actual [ἐνέργεια] and offered a couple of examples: the statue of Hermes is 'potentially [δυνάμει]' in the block of wood and the half line is 'potentially' in the whole line because they 'might be separated out'. As he put it, if the status of Hermes has been separated out (of the block of wood) and the half line has similarly been separated out (of the whole line), then they are being 'actually [ἐνεργείᾳ]', as opposed to potentially. Significantly, Aristotle adds another example: we call scientist [ἐπιστήμονα] someone who is not studying but who has the power to study ('καὶ ἐπιστήμονα καὶ τὸν μὴ θεωροῦντα, ἂν δυνατὸς ᾖ θεωρῆσαι'). Unlike the first two examples, which refer to substances, the third example refers to a power, to study. Still, Aristotle's point is that someone who has the power to Φ without actually Φ-ing is 'potentially Φ-ing'. Immediately after these three examples, Aristotle (1048a35–b5, 2:1655) offers more examples of the use of the actually/potentially dipole, aiming to suggest (by induction, as he says) a more systematic distinction:

that which is building to that which is capable of building;
the waking to the sleeping;
that which is seeing to that which has its eyes shut but has sight;
that which is shaped out of the matter to the matter;
that which has been wrought to the unwrought.

Of these contrasts, one member is 'actually' while the other is potentially. Still, he notes (my emphasis), there is no strict definition here, since 'all things are not said in the *same sense* to exist actually, but only by analogy…; for some are as movement to potentiality, and the others as substance to some sort of matter.' The examples then are not uniform; some refer to powers, some refer to substances. But the contrast is good enough to suggest that as we can say of something that has its eyes closed but has sight that it does not actually see though it potentially sees (that is, it has the power to see), so we can say of something that has the power to build but is not actually building that it builds potentially. What exactly does that mean? As Aristotle (1049a5–10, 2:1656) explains, X is potentially Φ-ed if X has the power to be Φ-ed and 'nothing in it' prevents it' from being Φ-ed or 'nothing external is interfering'.

Commenting on the broader sense of 'potential' that Aristotle tried to canvass in *Metaphysics* Book 9, David Ross (1924, cxxvi; see also Stein 2014 and Makin 2006) takes it that powers qua generators of change are transeunt powers; hence they relate two things. But potentiality, he thinks, is a 'capacity in A of passing into a new state of itself'. Hence, according to Ross, potentiality is not a transeunt power but an immanent power. The relation between the two then is this. A transeunt power is at the same time an immanent power, since when A produces a change in B 'A is itself passing from potentiality to actuality'. But immanent powers are not necessarily transeunt powers, since in this case A does not act on something distinct but it 'merely passes from a relatively unformed to a relatively formed condition, as when the wood which is potentially a statue become an actual statue'. But since transeunt powers are immanent powers too, a non-exercised power P is in *potentiality* and passes into actuality when it acts. This suggests that being in potentiality is a *mode of existence* which is contrasted to another mode of existence, corresponding to *actuality*.

For our purposes, it is important to stress that powers qua principles of change in something else are in two modes: they are potentially and they are actually. When they are potentially, they do not act; when they are actually, they are in act. But a power can be without acting; and for Aristotle (A 1049b5, 2:1657) the power is not unreal because of this. Even if 'actuality is prior to potentiality', potential being is a kind of real being. Michael Frede (1994, 184) noted that, insofar as a power exists without acting, it has a degree of reality: it is a potentiality as opposed to an actuality. As he put it, 'a basic δύναμις can be thought to confer some minimal degree of reality on what has it, indeed that having it is a way of being real, namely to be a potential, rather than an actual being.'

All this is relevant for the following reason. When they are not acting, powers exist potentially; hence, they can inhere in substances independently of the *actual* existence of other substances. Yet, they cannot be actual unless they are actualized and they cannot be actualized without other substances. Hence, Aristotle's view seems to be that the power of X to Φ is possessed potentially even if there is no Y with the passive power to be Φ-ed. But the presence of Y is a condition for the activation of a power of X to Φ and hence for its *actual* possession by X.

However, two perishable elements—earth and fire—are always in act. In this sense, they 'imitate' imperishable things which are only actual and not in potentiality for anything—they are in actuality (1050b23–5, 2:1659): the sun, the stars, and the heaven always act. Like them, earth and fire 'are ever active; for they have their movement of themselves and in themselves' (1050b29–30 2:1659).[5]

---

[5] At this point Aristotle (1050b31–5, 2:1659; translation amended) notes that all other potencies [δυνάμεις] 'are for the opposites; for that which can move another in this way can also move it not in this way, that is if it acts according to reason. But the same non-rational powers can produce opposite results only by their presence or absence.' This seems to contradict Aristotle's claim that natural powers can produce only one effect, according to their natures. For a discussion, see Makin 2006, 219–20. Part of the point Aristotle makes here is that powers which are not acting constantly may be prevented from acting, or their action may be interfered with.

4. The Relativity of Powers

Powers, then, have two modes of being: in potency and in act. But either way, being in potency and being in act, powers are towards something. Aristotle explicitly takes *power* (the active–passive pair, to be more precise) to be one of the three cases of the non-substantial category of being relative [*pros ti*]. In introducing the very category of being relative in *Metaphysics* Book 5 (A 1020a26–1021b11, 2:1612–13), and after he has introduced the primary sense of power noted above, he adds that a clear case of relatives are 'that which can heat to that which can be heated, and that which can cut to that which can be cut', and he adds, 'and in general the active to the passive'. Hence, powers are relative in a *dual* sense: when they are in a subject potentially and when they are being actualized. The agent and the patient are relative both in terms of their respective active and passive powers and in terms of the actualization of these powers. The example he immediately offers is revealing: 'that which has the power of heating is related to that which has the power of being heated, because it can heat it, and, again, that which is heating is related to that which is being heated and that which is cutting to that which is being cut, because they are actually doing these things.' Clearly, it is one thing to have the power to cut without actually cutting anything, and it is quite another thing to be cutting something. In the first case the power is possessed but is not activated; in the second case, it is possessed *and* activated. In either case, powers are relative in the sense that their ascription refers their bearer to something with the relevant passive powers.

In *Categories* Aristotle (6a37–6b11, 1:10–11) offered two definitions of relatives [*pros ti*]. On the first, 'We call *relatives* all such things as are said to be just what they are, *of* or *than* other things, or in some other way *in relation* to something else.' What's important in it is that a relative 'is said [λέγεται]' what it is 'in relation to something else'. Aristotle notes that this account leads to problems, the chief being that it allows substances (or parts thereof) to be relative—which is impossible, given his category of substance. He then settled for an ontic version of the definition above, viz. 'those things are relatives for which *being is the same as being somehow related to something*' (8a31–2, 1:13). This is a substantial move from *what is said* to *what is*: from λέγεσθαι to εἶναι. On this account, relatives hold *somehow toward something* [τῷ πρός τί πως ἔχειν]. Being relatives, then, have their being toward something.

The key characteristic of relatives is that they have correlatives [ἀντιστρέφοντα]. So, for a man to be a slave there must be another one who is his master, and conversely. One cannot be a slave, which is a relative attribution, if one is a solitary object in the universe. Similarly, for a man to be a father there must be another man or woman who is his son or daughter. So, if Φ is a relative attribute, it should be the case that X is Φ of Y iff Y is 1/Φ of X, where 1/Φ is Φ'. For example, X is a slave of Y iff Y is a master of X.

In *Metaphysics* Book 5, Aristotle (1021a27–8, 2:1612) noted that the relatives that are powers [δύναμιν] are called relative [*pros ti*] because 'their very essence includes in its nature a reference to something else, not because something else is related to *it*'.[6] Perhaps a better translation of the Greek passage would be that power is relative because 'when it comes to their being, they refer to something else'. So, positing powers implies that they 'point to' something outside themselves, viz. something that will suffer change when they act (having the suitable passive power) or something that will cause change in them (having the suitable active power). So powers are relative *irrespective* of whether or not they are in potency or in act. When in potency they are directed towards a possible correlative; where they are in act they are directed towards an actual correlative. Their relativity implies their towardness; and their towardness implies relatedness. As Aristotle put it in *Physics* (195b26–8, 1:334), 'powers are relative to possible effects, actually operating causes to things which are actually being effected.'

But powers belong also to the category of quality. A quality is 'that in virtue of which things are said to be qualified somehow' (8b25–6, 1:14). And what Aristotle calls natural power [δύναμις φυσική] are cases (one of the four) of quality: 'For it is not because one is in some condition that one is called anything of this sort, but because one has a natural power [δύναμιν φυσικήν] for doing something easily or for being unaffected' (9a19–21, 1:14; translation amended).

Natural powers then are both qualities *and* relatives. Aristotle thinks there is no problem in an attribute belonging to two genera, since 'if the same thing really is a qualification and a relative there is nothing absurd in its being counted in both the genera' (11a38–9, 1:17). Though the category of being relative is not the same as what we would today call a relation, since relatives are not 'shared' by two relata: being relative implies *relatedness*; what is relative holds *somehow toward something*. A relative attribution refers its subject to something outside it. Otherwise, it wouldn't be relative. But how can it be that a quality is such that its being is 'the same as being somehow related to something'?

---

[6] In this passage Aristotle distinguishes between the first two kinds of relatives and the third—the so-called intentional relatives such as knowledge and perception. But all this need not detain us here.

In the Aristotelian *Categories* anything which is not a substance is an attribute, and attributes inhere in substances. Being a slave, then, is a relative attribute which has to inhere in a substance (a person in this case) but it has its being towards something else (a master) since it cannot be possessed by a substance unless there is a correlative. Similarly, having the power to cut is a relative attribute which has to inhere in a substance (for example, a sharp knife) but has its being towards something else (that which can be cut).

It is precisely because powers are directed, and hence, they incorporate relatedness, that they can be principles of connection among distinct existences. But relatives are always treated by Aristotle as relational properties of a single bearer. Qua accidents, relatives can only inhere in one substance and never in two (at the same time). Powers, then, are relational properties of a substance *and* inherent to a substance. Aristotle doesn't explore this peculiar feature of powers very much but, as we shall see in the next section, it becomes central in understanding powers in the Middle Ages. This is the source of the idea that powers have *directedness*.[7]

5. The Directedness of Powers

Following Aristotle, Thomas Aquinas took powers to be qualities. Ultimately, powers are active and passive qualities of the elements (*ST* 1.115.2, 1:561).

Natural powers are natural tendencies [*inclinationes*] (see Geach 1961 for discussion). Objects have some natural tendencies and they lack others. As Aquinas put it, 'Thus it is clear that fire has a natural tendency to give forth heat, and to generate fire; whereas to generate flesh is beyond the natural power of fire; consequently, fire has no tendency thereto, except in so far as it is moved instrumentally by the nutritive soul' (*ST* 1.62.2, 1:305–6). Hence, each natural substance has its own natural powers (1a.6.5, 1:619). The difference between active and passive powers is a difference in their relations to their respective proper objects. In the case of passive powers, the object is to the act of passive power 'as the principle and moving cause'. In the case of active powers, the object is 'a term and end'. Hence, the object of a power is either the principle of the action (passive power) or the end of the action (active power). The passive power of being capable to be warm has as its object the being warm—which requires an active principle (a hot object); the

---

[7] For a classic account of Aristotle's view of relatives, see Weinberg 1965, 68–78. For a detailed study of Aristotelian relations in light of modern accounts, see Brower 2016.

active power of heating has as its object something which can become hot, which requires an end—making something warm (1.77.3, 1:385–6).

Powers can be assisted by other powers. But they can operate and bring about their effect only if the effect is not beyond the natural capacity of the power. Powers can be hindered by other powers. Hence, powers have limits in their operations, determined by their natural capacities and the hindrances of other powers. The natural capacity determines what a substance possessing a power can do; what is beyond the natural capacity is impossible—for example, a human being cannot fly. Or, the soul has the power to move in any direction, but is hindered by the weight of the body; hence a human (body plus soul) cannot 'mount upwards' (1.62.2, 1:305–6).

Powers have strengths. For a power Φ of X to bring about a change in Y, the passive power of Y should not exceed in strength the active power Φ of X (1.82.2, 1:414). Powers act of necessity (as long as the power is unimpaired) (1a.71.4, 1:899–900). Hence, powers act uniformly: 'the natural active principles are always determined to the same acts' (1a.72.3, 1:904).

A key feature of powers is that they have *directedness*. As Aquinas put it, 'A power as such is directed to an act' (1.77.3, 1:385–6). This feature of powers is grounded in their being relative. Following Aristotle, Aquinas (1961, 5.17.1002, 1:n.p.) explained in his *Commentary on Aristotle's Metaphysics* Δ that active and passive things are relative 'according to acting and undergoing, or to active and passive potency'.

Now, as Aquinas notes, a real relation consists in the bearing of one thing upon another. When it comes to powers (one of the three Aristotelian cases of relatives), X bears on Y 'according to active or passive power, inasmuch as one thing receives something from another or confers it upon the other' (5.17.1004, 1:n.p.).

The ascription of power then implies relatedness (that is, it is a relative attribution). In *De potentia Dei*, Aquinas (2012, 7.11, 225) notes that 'in order that some things have a relation, each needs to be a distinct being (since there is no relation of the same thing to itself) and be able to be related to something else.' But relation (being relative) signifies 'something in transit to something else, not something abiding in a subject' (7.8, 216–17). Hence: relation is a kind of predication according to which what is attributed to the subject is signified as passing from the subject of the relation to the terminus. Since no attribute can inhere in more than one substance, relatives have a subject (in which they inhere) and a terminus (to which they are *towards*). Relations cannot exist without foundations; that is, without inhering in some substance and being dependent on its qualities. Aquinas is clear that relative attribution or relative denomination does not imply composition between the subject and the terminus: 'what we attribute to something as proceeding from it into something else does not produce a composition of the other with it, just as an action does not produce a composition with an active thing' (7.8, 217).

6. The Duality of Powers

Like Aristotle, then, Aquinas took it that power, qua a mode of being, belongs to the category of relative [*pros ti*; *ad aliquid*]. In *Summa contra gentiles* he (Aquinas 1923, 2.10.1, 13) noted: 'power implies relation to something else as having the character of a principle (for active power is the principle of acting on something else, as Aristotle says in *Metaphysics V*).' But powers are qualities too (2.8.5, 11–12). Hence, they have to inhere in a substance. For as he put it in the *Summa theologica* 'the essence of an accident is to inhere' (*ST* 1.28.2, 1:152).

To address this issue, in his *Commentary on Aristotle's Metaphysics Δ* Aquinas (1961, 5.9.892, 1:n.p.) distinguished between three ways in which the accident is *in* the subject:

- essentially and absolutely as something following from its matter, corresponding to the category of quantity;
- essentially and absolutely as something following from its form, corresponding to the category of quality;
- not present in the subject absolutely 'but with reference to something else', corresponding to the category of relation.

For Aquinas, not all qualities are powers (see Kahm 2016 for discussion). As he says, 'the term white, as it is used in the categories, signifies quality alone' (5.9.894, 1:n.p.). But powers are qualities. An example would be gravity. Bodies have an 'inclination and order to the centre'. Hence, heavy bodies are such that, by nature, there exists an inclination in them towards the centre (and 'the same applies to other things'). In cases such as these, where the quality possessed by a subject is a power (a natural tendency; an inclination to something else), it is proper to say that the power (qua quality) inheres in its subject, but (qua relation) it 'signifies only what refers to another'. Hence powers qua qualities are *inesse* and qua relations are *ad aliud* (*ST* 1.28.1, 1:151–2).

In fact, it is this natural inclination of powers to bear on something else that makes powers *real* relatives. As Aquinas (1961, 5.17.1005, 1:n.p.) put it, a quality inheres in a substance and a thing can bear on another thing 'only inasmuch as quality has the character of an active or passive power, which is a principle of action or of being acted upon'. But if powers are qualities of substances then powers qua qualities are in the subject essentially and absolutely and qua relatives are in the subject not absolutely 'but with reference to something else'. The relativity of powers grounds their directedness; the qualitative character grounds their inherence.

The key point here is that this dual aspect of powers grounds their two main features: inherence and directedness. As Aquinas (*ST* 1.28.2, 1:152–3), following Aristotle, noted: relation qua a category of being has a dual character. On the one hand, it is one of the nine categories of accidents and hence it has to inhere in a subject, since the essence of an accident is to inhere [*accidentis enim esse est inesse*]. On the other hand, each category of accident has its own 'proper nature'; its own *propria ratio*. Quantity has as its ratio the measure of substance, while quality is the dispositions of substance. Relation has as its *propria ratio* its being toward something outside the subject [*ad aliquid extra*]. Hence, relations 'signify a respect which affects a thing related and tends from that thing to something else; whereas, if relation is considered as an accident, it inheres in a subject, and has an accidental existence in it. Gilbert de la Porree considered relation in the former mode only'. Understanding relatives is imperative since it is via them that there is natural order in the world, viz. it is via relatives that the various parts of nature are related to each other in various ways. Hence, not only is it the case that the things outside the mind are, quantitatively and qualitatively, in a certain way; they are also related to each other in certain ways. This for Aquinas contributes to the 'perfection and goodness in things outside the soul'. Hence, 'there need to be some relations in things themselves, by which one thing is related to another'. This very ordering is dual: 'one thing is ordered to another thing by quantity or by an active or passive power, since we note something in one thing in relation to something extrinsic only by these two things' (2012, 7.9, 220). To put it bluntly, the natural order in the world is either quantitative or causal.

As he explains further in *De potentia Dei* (7.9, 221), the order (relation) between any two substances requires that substances have qualities which inhere in them as accidents. But if we think of the order *itself* qua relation, then its sole being is its towardness [*ad aliud*], since in this case we conceive of the order 'as if passing into the other and somehow supporting the related things'. This action is considered 'to issue from an active thing, but, insofar as it is an accident, we consider it as in the active subject'. If we consider the *order* between two substances (for example, the sameness in colour between Socrates and Plato, the relation between father and son, or the relation between fire and wood), an accident which is taken to be relative is such that 'its essence is not perfected as it exists in its subject but it perfected as it passes into something else'.

Hence, understanding natural order requires thinking of relatives as 'mere towardness [*sed solum quod ad aliud*]'. A relative, that is a relative denomination, is inherent to the subject insofar as it is not a relation; and it is a relation ('mere towardness') insofar as it is not inherent to the subject; action, for example, is a relation insofar as it is issued *from* the agent *towards* the terminus but it is an accident insofar as it is inherent in, that is, it is based on a quality of, the active agent. As Aquinas (7.9, 221) states, if the terminus is removed, the essence of the accident (the active power) is removed as regards its action but remains as regards its cause.

An objection to the towardness of relations was that a relative can come to be without any intrinsic change in the subject; and it can also be lost without an intrinsic change in the subject. A relation of X to Y can cease to be without any change in X, if Y is changed. If a father loses his son, he ceases to be a father, without changing intrinsically. Similarly, X can acquire a new relation (to Y) if Y changes. For instances, Socrates acquires the relative being-similar-in-height-to-Plato by Plato becoming as tall as Socrates.

If relatives can be acquired or lost without any change in the subject that possesses them, in what sense are they *in* the subject? Relations require a foundation in the subject. Hence, if the terminus is removed, the foundation remains but there is no longer a relation in the subject. For instance, if a father loses his son, he ceases to be a father, but he can still retain the power of begetting, which is the foundation of his relative accident (*ST* 1.40.4, 1:207). But what happens when a subject acquires a new relative, for example when Socrates becomes equal in height to Plato by Plato becoming as tall as Socrates? The subject is not really changed; and yet the subject acquires something, viz. a relative predication. Aquinas's answer to this is based on the metaphor of the root. He says:

> Hence it must be said that if someone by changing becomes equal to me, and I do not change, then that equality was first in me in some way as in a root, from which it has real being. For since I have a certain quantity, it happens that I am equal to all those who have the same quantity. When, therefore, something newly takes on that quantity, this common root of equality

is determined in regard to him. Therefore, nothing new happens to me because of the fact that I begin to be equal to another, because of his change.

(1999, 5.3.667, 327)

This passage is perplexing (for discussion, see Henninger 1987; Henninger 1989; Mugnai 2016; Penner 2016). Aquinas states that the relative is in the subject 'in some way as in a root'. Because of this, when Y acquires a property F that makes him equal in height with X, X acquires a relative R (equality-in-height-to-Y) without any intrinsic change in him because X has a property F ('the common root of equality'), which gives real existence to R. It is because of F that X *already* has R in him 'in as in a root'. A natural way to read Aquinas's root metaphor is this:

X has height F;
F in X is the root of 'being-equal-in-height-with-whoever-has-height-F';
because of F, X can have the relative *equality-in-height*;
Y has height F;
hence, X has the relative *equality-in-height-with-Y*.

The key point here is that, when there is no actual terminus Y, there is only the foundation for the relation in X. When the terminus is posited, the relation acquires real existence *from the foundation*, that is from the actual quality that is its foundation.

What's the difference, then, between equality vis-à-vis a certain quality (for example, whiteness) and power, though both are relatives? Whereas the foundation of equality-in-colour *qua* relative is the quality (whiteness), the foundation of power qua relative can be the power itself qua quality. Here is how Aquinas (2012, 7.9, 220) put it: 'Each thing acts upon another by its active power, and another acts upon it by reason of its passive power. But substance orders something only to itself, not to something else except incidentally, namely insofar as a quality, a substantial form, or the matter has the aspect of an active or passive power, and insofar as we consider an aspect of quantity in it.'

There is no reason to think that all powers are rooted in themselves, since complex powers might well be rooted in other qualities. Fragility or solubility—to use modern examples—need not ground themselves since they are 'complex' powers. To call a substance 'fragile', that is to say, to attribute to it the passive power to shatter, is to make a relative attribution whose foundation is other qualities, some of which might also be powers of the substance. But the basic powers (the powers of the four elements) are 'rooted' in themselves. In this sense, at least some powers, qua qualities *and* relatives, are their own foundations. This peculiarity of power has to do with the fact that powers confer on the substances that possess them inclinations towards acting on other substances. As Aquinas (*ST* 1.60.4, 1:300) put it, 'fire has a natural inclination to communicate its form [heat] to another thing.'[8]

Now, Aquinas (1961, 9.5.1825–9, 2:252–4) agreed with Aristotle that powers can be in two states: potentially and actually. As he explains in his *Commentary on Metaphysics Book* Θ, 'a thing is actual when it exists but not in the way in which it exists when it is potential.' He takes Aristotle to have offered two senses of actuality, the first of which 'means action, or operation'. Hence, one who has the power to contemplate has this power 'even though he is not actually doing so'. But the power exists potentially. Whereas when he is contemplating—that is, when the power is actualized—the power is 'in a state of actuality'. This distinction between two states of a power can be captured, in general, by 'the relation of motion to motive power or of any operation to an operative potency'. Only fire and earth among substances are such that their inherent powers are always acting 'inasmuch...as their forms are principles' of their characteristic manifestations (9.9.1880, 2:273).

What does it mean then to say that some powers ground themselves? I take it to imply that the foundation of a power qua relative is the very same power in potentiality. When the power is actualized it enters into actual and real relations with the correlative powers. But because the power exists potentially too, the actualization, and hence the real relation, was in a sense in the substance 'in some way as in a root'. When a power shifts from potentiality to actuality there is a real relation 'inasmuch as one thing receives something from another or confers it upon the other' (5.17.1004, 1:436). But it should be stressed that, even if powers have their roots in themselves, they have the dual character of being qualities, and hence being inherently possessed by their bearers, *and* relatives, and hence having a being toward something else.

7. The Distinction between *Inesse* and *Esse-ad*

This issue of the dual status of powers was intimately connected with the problem of reality and distinctness of relatives. The later scholastic

---

[8] Pasnau (2011, 519) has noted that medieval powers are taken to be causally active categorically and not conditionally. They have 'intrinsic rather than derivative causal powers'.

philosopher Francisco Suárez (*DM* 47.1.8, 2:784a; 2006, 46) summarized the various positions in this debate and noted that the 'received opinion' had been that in created things 'there are real relations, which constitute a genuine and special category'. But being real does not imply that relations are distinct from their foundations. The distinction between the relation and its foundation was taken to be conceptual or one of reason as opposed to being real. For Suárez and the received opinion, relations are not free-floaters. A relation 'is not a thing that has in itself an entity that is distinct from all absolute entities' (47.2.2, 2:783b; 2006, 52). The key argument for this view was that a real distinction would require a two-way separability of the relation from its foundation. But 'even though the foundation can remain without the relation, still the relation cannot in any way remain without the foundation' (47.2.5, 2:786b; 2006, 53).

The being of a relation, then, is not in reality distinct from the being of its foundation; the distinction between the two is a *conceptual* distinction. Still, the relation entails some real and intrinsic denomination of a proper relative thing: 'But it must be understood that relation indeed entails some form that is real and intrinsically denominating a proper relative thing (*relativum*), which [relative thing] it constitutes [as relative]' (47.2.3, x:xxx; 2006, 73). Hence, relatives are real but not really distinct from their foundations. A relative is an absolute form, not taken absolutely but as 'respecting another [form], which the relative denomination includes or connotes' (47.2.22, 2:792b; 2006, 74).

To claim this, in other words, is to claim that relations have towardness [*esse ad*]. Repeating Aquinas's views, Suárez (47.5.4, 2:806a; 2006, 109) took it that relation is an accident 'whose total being is being toward another'. But this is to acknowledge the main puzzle: how can it be that the total being of a relation is towards another if it is an accident? Being an accident, a relation must have something of its being in a subject. Hence it cannot have its total being consist 'in a disposition towards another'.[9]

Suárez's solution is no different from Aquinas's: the being toward, the real and proper being of a relation, has to be a being in because otherwise the relative attribution would be nothing real in the nature of things. Hence a categorical relation would not be something real. Still, a real relation does not have absolute being; it indicates 'that the being of a relation as such does not

---

[9] 'But then there arises a second difficulty.... For if a relation is an accident, its whole being cannot then consist in a disposition toward another. For it is necessary that something of its being be in a subject, so that in that way it can be an accident, since the being [*esse*] of an accident is "being in" [*inesse*].' DM 47.5.6, 2:806a; 2006, 110.

stay in the subject that in its own way it affects and denominates but rather it orders that [subject] to a terminus, and in this is placed the whole formal character of a relation' (47.5.6, 2:807a; 2006, 111).

When it comes to relations, then, the distinction between *inesse* and *esse-ad* is conceptual. Among qualities, 'a potency entails an essential relation to its object'. But potencies are qualities too. Take gravity, which is a quality of things in virtue of which they have an inclination to fall down. The *esse-ad* and the *inesse* are only conceptually distinct: 'by affecting and informing the heavy thing itself [gravity] inclines that thing, and this is to relate [*referre*] it toward the center [of the universe]' (47.4.8, 2:801a; 2006, 95). The *esse-ad* cannot be prescinded from 'being in'. Suárez's point was that, though a 'real relation does not add [another] thing to absolute things or a real mode that is really [*ex natura rei*] distinct from those [absolute things]', it does not follow that 'a relation of this kind is completely nothing' (47.2.24, 2:793b; 2006, 76–7).

8. Powers: Real but Not Distinct

Let's take stock. According to the Aristotelian–Thomist account of powers, power is the principle whereby one thing acts on another; hence, powers are principles of connection among distinct substances. Powers qua qualities are inherent in substances; but they also ground the order of the world by having their being towards something else—the *substances* which have the passive powers required for their actualization. The very same attribute qualifies its bearer *and* points towards something else.

I have claimed that Aristotelian–Thomist powers are best seen as having a dual aspect, being both what we would nowadays call qualitative (categorical) and dispositional. But thinking of powers as having a dual aspect requires thinking of them as being *relative*. Given the Aristotelian–Thomist account of relatives, the relativity of powers does not imply that they are somehow in between the subject and the terminus. They are in the subject. Besides, the relativity does not imply that powers are not real. But it implies that they are not really distinct from their qualitative basis, since the *esse-ad* cannot exist without the *esse-in*. Hence, the distinction between the two ways to view a power—the qualitative and the directed—is conceptual.

I also argued that the basic or fundamental powers, but not the derivative powers, ground themselves: qua relative, they are their own foundation. But they do so precisely because they are not 'pure powers', pure '*esse-ad*', but qualities too. Still, the distinction between *inesse* and *esse-ad* is an important distinction to draw because it highlights how a certain property can both qualify its bearer *and* be such that it, potentially or actually, relates its bearer to some other object. Hence, Aristotelian–Thomist powers are not what nowadays may be called 'pure powers'; they are not such that they have their full being purely towards something else. They are determinations of objects (substances) which are *both* qualities and relatives. They are qualities insofar as they inhere in substances and contribute to their qualification; they are relatives insofar as they are conceived of as endowing the substance that bears them with a potential to relate to other substances.[10]

Hence, a property is not characterized by 'mere directedness' simply because, if it were so characterized, it would be nothing but the possibility of an actualization, should the opportunity arise. As Mumford (2003, 15) puts the point, 'If such dispositions were unmanifested, it would appear that the particular would have no manifest properties—nothing displayed—and any particular with no manifest properties seems like nothing at all.' And though Mumford takes it that this might be a price the friends of powers have to pay if they are to accept ungrounded powers, paying it can be avoided if we take it that powers are qualities too; that is, that a power has a qualitative nature in virtue of which it inheres in substances even when it is not acting. The directedness of powers, then, is only conceptually distinct from the qualitative nature of the property.

This view, it should be stressed, is akin to the position advocated by Heil (2003, 11) and C. B. Martin (1994): the so-called 'identity theory' of properties. According to this view, 'every property of a concrete spatiotemporal object is simultaneously qualitative and dispositional. A property's "qualitativity" is strictly identical with its dispositionality, and these are—are strictly identical with—the property itself' (Heil 2003, 111). The chief difference, I think, is that on the Aristotelian–Thomist view, not all qualities are powers, though qualities, such as being white, are the foundations of the relational properties their bearers might come to have.

An important advantage of the Aristotelian–Thomist view is precisely that in distinguishing between two modes of being, that is, being in potency [*esse potentia*] and being in act [*esse actu*], it becomes conceptually possible to think of powers as being in potency when they are not exercised and as such to inhere in their bearers qua qualitative attributes of their bearers. It is by virtue of these attributes that substances can be related to, and act upon,

---

[10] For a similar view see David Peroutka 2012. I disagree with Peroutka's characterization of powers as '*relatio transcendentalis*,' but this issue need not detain us here.

others; and this is what it is for these attributes to have their being towards something else.

Powers, then, are real, but not really distinct from their qualitative grounds. Their directedness is the potential by which, qua relatives, powers endow their bearers to enter into relations with other substances. If there were a lonely object in the world, the potential for relations would be there, since the object would have some inherent qualitative properties but no relatedness, and hence no change. But since there are various entities (substances) in the world, powers are actualized and this accounts for the natural order in the world, viz. for substances being related to each other.[11]

References

Aquinas, Thomas. 1923. *Summa contra gentiles*. Trans. Fathers of the English Dominican Province. London: Burns, Oates & Washbourne Ltd.

Aquinas, Thomas. 1961. *Commentary on the Metaphysics of Aristotle*. Trans. John P Rowan. 2 vols. Chicago: Henry Regnery Company.

Aquinas, Thomas. 1999. *Commentary on Aristotle's Physics*. Trans. Richard J. Blackwell, Richard J. Spath, and W. Edmund Thirlkel. Notre Dame, IN: Dumb Ox Books.

Aquinas, Thomas. 2012. *The Power of God [De potentia Dei]*. Trans. Richard J Regan. Oxford: Oxford University Press.

Brower, Jeffrey E. 2016. 'Aristotelian vs Contemporary Perspectives on Relations'. In Marmodoro and Yates 2016, 36–54.

Frede, Michael. 1994. *Essays in Ancient Philosophy*. Minneapolis: University of Minnesota Press.

Geach, P. T. 1961. 'Aquinas'. In *Three Philosophers*, ed. G. E. M. Anscombe and P. T. Geach, 56–125. Oxford: Basil Blackwell.

Heil, John. 2003. *From an Ontological Point of View*. Oxford: Clarendon Press.

Heil, John. 2005. 'Dispositions'. *Synthese* 144, no. 3 (April): 343–56.

Henninger, Mark Gerald. 1987. 'Aquinas on the Ontological Status of Relations'. *Journal of the History of Philosophy* 25, no. 4 (October): 491–515.

Henninger, Mark Gerald. 1989. *Relations: Medieval Theories, 1250–1325*. Oxford: Oxford University Press.

Irwin, Terence. 1988. *Aristotle's First Principles*. Oxford: Clarendon Press.

---

[11] Many thanks are due to Stavros Ioannidis, Henrik Lagerlund, Michalis Philippou, and Vassilis Sakellariou for incisive comments on earlier drafts. Many of the ideas of this chapter were tried in graduate seminars in the University of Athens and the University of Western Ontario.

Kahm, Nicholas. 2016. 'Aquinas on Quality'. *British Journal for the History of Philosophy* 24 (1): 23–44.

Lowe, Jonathan E. 2016. 'There are (Probably) No Relations'. In Marmodoro and Yates, 2016, 100–12.

Makin, Stephen, trans. 2006. *Metaphysics Book* Θ by Aristotle. Oxford: Clarendon Press.

Marmodoro, Anna and David Yates, eds. 2016. *The Metaphysics of Relations*. Oxford: Oxford University Press.

Martin, C. B. 1994. 'Dispositions and Conditionals'. *The Philosophical Quarterly*, 44, no. 174 (January): 1–8.

Molnar, George. 2003. *Powers*. Oxford: Oxford University Press.

Mugnai, Massimo 2016. 'Ontology and Logic: The Case of Scholastic and Late-Scholastic Theory of Relations'. *British Journal for the History of Philosophy* 24 (3): 532–53.

Mumford, Stephen. 2003. Introduction to *Powers* by George Molnar. Oxford: Oxford University Press.

Mumford, Stephen, and Rani Lill Anjum. 2011. *Getting Causes from Powers*. Oxford: Oxford University Press.

Owen, G. E. L. 1965. 'Inherence'. *Phronesis* 10, no. 1 (January): 97–105.

Pasnau, Robert. 2011. *Metaphysical Themes 1274–1671*. Oxford: Clarendon Press.

Pechlivanidi, Elina, and Stathis Psillos. 2020. 'What Powers are Not'. In *Dispositionalism: Perspectives from Metaphysics and the Philosophy of Science*, ed. Anne Sophie Meincke, Synthese Library 417, 131–150. Dordrecht: Springer.

Penner, Sydney. 2016. 'Why Do Medieval Philosophers Reject Polyadic Accidents?' In Marmodoro and Yates 2016, 55–79.

Peroutka, David. 2012. 'Dispositional Necessity and Ontological Possibility'. In *Metaphysics: Aristotelian, Scholastic, Analytic*, ed. Lukáš Novák, Daniel D. Novotný, Prokop Sousedík, and David Svoboda, 195–208. Berlin: Walter de Gruyter.

Ross, W. D. 1924. *Aristotle's Metaphysics. A Revised Text with Introduction and Commentary*. Oxford: Clarendon Press.

Stein, Nathanael. 2014. 'Immanent and Transeunt Potentiality'. *Journal of the History of Philosophy* 52, no. 1 (January): 33–60.

Suárez, Francisco. 2006. *On Real Relation (Disputatio Metaphysica XLVII)*. Trans. John P. Doyle. Milwaukee: Marquette University Press.

Weinberg, J. R. 1965. *Abstraction, Relation and Induction: Three Essays in the History of Thought*. Madison: University of Wisconsin Press.

Yates, David. 2016. 'Is Powerful Causation an Internal Relation?' In Marmodoro and Yates, 138–56.

2

Powers, Possibilities, and Time

Notes for a Programme

Calvin G. Normore

We deliberate about the future but not the past or the immediate present. We strive to increase our abilities, capacities, and powers to enable us to bring about what we choose and not what we take to be already or have been the case. These truisms suggest that there are connections among possibilities, powers, and time. What are those connections?

The earliest reflective discussion of modality of which we have evidence, the 'Megarian' picture crystallized in the Master Argument of Diodorus Cronus, attempts to explicate modality in terms of time and truth. We know from Boethius, Epictetus, and Cicero that the Master Argument was a tri-lemma whose premises were:

1) What has been, necessarily has been.
2) From the possible, the impossible does not follow.
3) Something is possible, which neither is nor will be.

We are not told the argument for the incompatibility of the three, and we are not told of anyone who claimed the argument to be invalid. We are told that Diodorus himself rejected the third premise. Suppose we follow him. What emerges is a picture in which what is necessary is what always is, and what is possible is what is at some time or other. Reading the 'what' expansively, on this picture every genuine possibility is realized at some time or other.

The picture has a very respectable pedigree. A number of historians of philosophy, notably Jaakko Hintikka (1973), have suggested that Aristotle had it in mind. Nicholas Rescher (1963) has suggested it was a common view in medieval Islam. And Simo Knuuttila (1993) has suggested that it is a view we sometimes find in the Latin West, especially in early physics texts.

Calvin G. Normore, *Powers, Possibilities, and Time: Notes for a Programme* In: *Reconsidering Causal Powers: Historical and Conceptual Perspectives*. Edited by: Benjamin Hill, Henrik Lagerlund, and Stathis Psillos, Oxford University Press (2021). © Calvin G. Normore.
DOI: 10.1093/oso/9780198869528.003.0003

Perhaps even more tellingly, the statistical picture gets support from a striking development in formal semantics during the second half of the twentieth century, one apparently inspired by Leibniz—the treatment of both time and modality as forms of generality and of central temporal and modal notions as, or as best represented semantically by, quantifiers ranging over the members of a set of indices. This set is structured by what is conventionally called an accessibility relation, the idea being that what is the case relative to some indices is temporally or modally related to some others but perhaps not to all. From this perspective time and modality differ formally, at most, in the structural conditions placed on the accessibility relation.

I say 'at most' because it is unclear just how different temporal and modal logics need be formally to capture the corresponding notions. On the one hand, Gödel has shown us that in a model of general relativity time might be closed, so that any temporal situation might be accessible from any other. But, on the other hand, such metaphysical principles as the 'necessity of origin' suggest that the accessibility relation for modality must be at least a partial ordering. One might well wonder whether further reflection would show that formally modal and temporal logics do not differ at all!

Whether temporal and modal logics need differ formally has an analogue in the question whether the indices over which the quantifiers that express or replace modal operators need differ from those that express or replace temporal operators. Might times and (say) possible situations just be the same thing? Suppose we started with world stages. We would need enough for times. But would we need more?

At first glance, the identification of possibility with truth at a time is very implausible. Surely, we can think of alternatives to the way things were, and are, and will be! Despite the history and formal analogies then, Diodorus and those who follow him must make plausible the denial that such alternatives are possibilities.

We might start by noting that conceivability—at least the sort of conceivability we attribute to such alternatives—is no guide to possibility. Thomas Hobbes thought he could square the circle, but he was wrong; the circle cannot be squared. Why think that the supposed alternatives to the way things are, were, or will be, are genuine alternatives, that is possibilities?

1. Scotus, Ockham, and Auriol's Principle

Philosophical orthodoxy about possibility has varied over time. Medieval theorists commonly claimed that God can do anything that does not involve a contradiction. It is often unclear, however, what they thought the connection between conceivability and contradiction to be. Descartes famously claimed that God could bring about anything we could clearly and distinctly conceive, but he also claimed that the necessity of the past was itself clear and distinct. Spinoza and the absolute idealists of the nineteenth century insisted that only what was actual could be coherently conceived. Within the early analytic tradition, the common position seems to have been to take the notion of consistency formally—'contradiction' meant syntactic contradiction and 'entail' meant entail in some formal logic.

Such approaches have their own problems. Gödel showed us that, for any consistent and sufficiently strong theory, there are truths of arithmetic, and hence one supposes necessary truths, whose negations cannot be proved contradictory within that theory. Thus, if our notion of contradiction is formal, there are impossibilities that are not contradictory. If, on the other hand, we think of contradiction as a matter of the semantics of natural language, then, given our partial grasp of many (most?) natural language concepts, we may have little idea of what will prove to be possible. Is it possible that there is an Abrahamic God? Is it necessary? The semantics of natural language seems to give little guidance here.

More fundamentally perhaps, to identify possibility and conceivability seems to amount to a form of idealism—as if the limits of how the world might be were determined by the limits of how we could think about it. If there be a God, perhaps that would be true of it, but it seems unlikely to be true of us! In any case, the history of philosophy seems to have passed such approaches by. The second half of the twentieth century has seen a reaction against the conflation of alethic and epistemic modalities spearheaded by Saul Kripke's (1980) arguments that there were a posteriori necessities (such as, that atoms of gold necessarily had seventy-nine protons) and that the origins of things were essential to them (so that the current Queen of Canada could not have had different parents). The resulting picture of necessity makes more plausible the thought that many of the alternatives we conceive or imagine to be possibilities may not really be so.

Which, then, are the genuine possibilities? That would seem to be, at least in part, a matter of which restrictive metaphysical principles are true. Suppose, for example, that a Principle of Sufficient Reason is true, that causation is a matter of supplying sufficient conditions, and that explanation is typically causal explanation. Then it is impossible that a given situation obtain and there not obtain situations sufficient to bring it about. Or suppose true a Principle of the Necessity of Origins, which has it that a thing or state of affairs could not have had other causes than the ones it had. Then it is impossible that there be a thing and there not have been the things from which it originated. Both of these principles link modality and time; they constrain the past of any supposed present and, taken together, they entail that a present necessarily has a unique past.

What then of the present itself? Are there principles which constrain what might now be present? Can what is right now the case not be, right now, the case?

This was the issue on which two of the greatest modal theorists, John Duns Scotus and William Ockham, parted company. Scotus (1963, d.38, pt.2; d.39, q.1–5, n16, 417–18) claimed that in some cases we had what he called a non-evident (*non manifesta*) power at time t to do at t what we do not do at t. No such power was ever exercised but they were nonetheless real, and their reality was enough to make it true at the time that we do x that we can refrain from x. Ockham (1978, q.3, 533–4) denied that there were any such powers. He insisted that what is possible is possibly actual and drew as a consequence that the necessity of the past entailed that of the present.

Suppose we grant that the past is necessary. Since I have been on earth writing for the past few hours it is necessary that I was not flying to the moon over the past few hours. Given that right up to and including now I have been on earth, can I nonetheless be on the moon right now? Scotus thinks that in some very special cases something like this must be so. One such case was God's creative activity; God, at the moment of creation, was able to be not creating. Another was our free choices; at the moment of choosing to do the better we nonetheless have the ability to choose the worse. Scotus is, moreover, committed to the view that the existence of the relevant power at t entails the possibility of the state for which it is a power at t and so to the conclusion that God could right now not be sustaining the world and I could right now be choosing what actually I am not. Ockham disagrees. He takes it that a necessary condition for the existence of an ability or power is that it be realizable and so possibly actualized. He takes it too that not even God can simultaneously realize abilities whose realization is contrary or contradictory, and he takes it that (say) for me to realize my ability to do the better at the moment when I am doing the worse would require a change on my part. That is, Ockham accepts, and Scotus rejects, what I have called Auriol's Principle (Peter Auriol, I *Sent.*, d. 39): If A is F and A can be not-F, then A can change from F to not-F. Scotus rejects Auriol's Principle because he takes it that there is real possibility just in case there is the power to produce the relevant state and he takes it that, while the exercise of a power requires a change (from the merely potential to the actual state), the existence of a power does not require its actual realizability. Hence while A is actually F, it simultaneously can be not-F because there can coexist with the power for F-ing a power for not-F-ing even if that latter power is 'non-evident' and actually unrealizable.[1] Ockham, on the other hand, while agreeing that there is real possibility just in case there are the relevant powers, thinks the idea of an actually unrealizable power incoherent and, hence, rejects non-evident capacities and accepts Auriol's Principle.

What can we make of this disagreement? Within what by now is 'classical' modal logic there is no privileged position given to the actual world, and within 'classical' tense logic there is no privileged position given to the present time. This is reflected in the fact that the two most common types of model structure for such logics—in one of which there is an index taken as the 'actual' world or the present time, and in the other of which there is not—are provably equivalent. To capture a difference between them we need to extend the syntax of classical modal logic. The usual way of doing this is to add an operator—in modal logic an 'Actuality' operator and in tense logic a 'Now' operator. In both cases, the standard semantics for that operator has it 'snap' the sentence back to the privileged index so that no matter in what world or at what time we evaluate a sentence p we utter here and now, we evaluate A(ctually) p in the world of the utterance and N(ow) p at the present moment. Thus, the privileged world semantics for the Actuality operator does not validate L(p>Ap) but does validate Ap>Lap and that for the Now operator does not validate O(p>Np) (where O is an operator meaning something like 'at every time') but does validate Np>ONp. This contrasts with a natural reading of these operators in the semantics without a privileged world, which has it that p>Ap is true at each world and so L(p>Ap) and O(p>Np) are validated.

Obviously, a semantics which does not privilege the actual world or the present time has no resources to capture intuitions about the necessity of either, but the semantics for standard modal and tense logics, which gives the actual or the present a privileged position, goes too far (it would seem) in failing to distinguish two relations a sentence might bear to an index such as a world or a time. In the case of the tenses, Ockham (1978, q.3, 515 and 525) calls these two relations being about a time *secundum vocem* and being about that time *secundum rem*. We might think of the first as (merely) being *assigned true* relative to the index and the second as *being grounded* in the index. To

---

[1] For a fuller discussion of Scotus's relation to Auriol's Principle and relevant texts, see Normore 2003, 130–7.

see the difference, consider Ockham's account of future contingent sentences such as 'Peter is predestinate'. That sentence, says Ockham, is true right now (that is assigned truth relative to the present), but it is made true by what happens to Peter in the future—at the Last Judgement. The fact in virtue of which 'Peter is predestinate' is now true is not a present fact but a future fact. Similarly, while 'It is possible that Satan is predestinate' is assigned truth relative to our world, on the standard semantic picture it is true not in virtue of anything about our world (assuming that Satan is not going to be saved here) but in virtue of facts about other worlds, namely that he is saved in them. With this distinction in mind, we can distinguish two Actuality operators A_v and A_r and two Now operators N_v and N_r. Which ones capture what is meant by 'actually' and 'now'?

Debate still rages even about whether the English word 'actual' is best understood as relative to a context of utterance or to a context of evaluation. Let me be dogmatic. What is actually the case is what is settled in the actual situation, and the actual situation is the one you, gentle reader, are in. What is now the case is what is settled right now in the actual situation. From this perspective, it is plausible that $A_r p > L\ A_r p$ and $N_r p > L\ N_r p$ and plausible that there are not actually any alternatives to what is actual or presently any alternatives to what is present. From this perspective, it is not plausible that universally $p > A_r p$. If 'p' describes a situation that, for example, has not yet come to pass, it may be that it will be that p but it is not yet actual that it will be that p.

Suppose then that we accept Ockham's distinction between the sentences true at a time (*secundum vocem*) and those whose truth is, in my terminology grounded in the time (*secundum rem*). Suppose we assimilate this to the distinction between sentences to which we can and those to which we cannot prefix the operator A_r. What consequences does this have for what we take to be possible?

2. Modality and Time

So far, our focus has been on possibility. What, though, of time? Despite the thought experiments of Sidney Shoemaker (1969) and others and the effort of Newton in the Queries to his *Optics* and in the General Scholium to the second edition of *Principia* to consider space and time as the sensorium of God, there is a persistent intuition that time is connected with change. This intuition was presented forcefully, if a bit obscurely, by Aristotle (A 201a9-11, 1:343; 222b30–223a15, 1:377), who claimed that time is the measure of change and that change [*kinesis*] is in turn the actualization of powers qua powers [*dunameis*]. Putting the two claims together we get that time just is the measure of the actualization of powers or potentialities. This in turn suggests that there is something very odd about the idea of power over the past. If we can assume for a moment that there is a single time, we can see why. It is because the ordering of time simply follows the order of the actualization of powers. A power over the past would then be a power to make the past the future! The power view makes very natural the idea that the past is somehow fixed in a way that the future is not fixed.

The contemporary successor to Aristotle's view is the causal theory of time, formulated in its modern forms by Reichenbach (1958), Grünbaum (1963), and Van Fraassen (1966). This approach comes in two flavours, strong and weak. The strong form argues that temporal phenomena can be reduced to causal phenomena—for example, that in the base case what it is for B to be later than A is for it to be that A does or could cause B; for B to be earlier than A is for it to be the case that B does or could cause A; and for B to be simultaneous with A is for neither to cause or be able to cause the other. The weak form does not claim reduction but rather that the temporal relations hold just in case and because the causal relations do. It seems likely that Aristotle maintained the strong version, arguing that we could not conceive of time without change and that the direction of time was as a matter of conceptual necessity given by the priority relations between causes and their effects. Sydney Shoemaker (1969) has argued, persuasively, that we can conceive of time without change, and although a number of puzzles remain, it is widely thought that temporally backwards causation is not inconceivable either. Still neither the conceivability of time travel nor Shoemaker's arguments speak to the claim that as a matter of metaphysical necessity temporal relations are parasitic upon causal relations. Suppose they do! Then if causality is a matter of the realization of powers and capacities, temporal relations, and in particular the 'arrow of time', will be a function of the fact that powers are metaphysically prior to their realizations.

What then of the relation between causality and the realization of powers and capacities? Might it be that A causes B just in case A has a power such that in the circumstances at hand B is the realization of that power? In *Causal Powers* (Harré and Madden 1975), just such a picture is suggested. On their account, a thing (what they term a 'powerful particular') has a nature in virtue of which in appropriate circumstances it will produce a determinate effect. Causality, they suggest, is just the activity of such powerful particulars.

If we take on board a causal theory of time and a picture of causation as the actualization of powers, we get, for what is actual, support for Auriol's Principle. If we take 'change' in Auriol's Principle to translate Aristotle's '*kinesis*', then the principle is powerful indeed. If true, it would entail the necessity of the actual past because no realization of a power is prior in time to its effect, and hence to the extent that possibilities are the realizations of powers in appropriate circumstances, there are no non-actualized possibilities with respect to the past. Moreover, it would entail the necessity of the actual present, because if time is the measure of the actualization of powers, then the actualization of a power takes time and so no realization of a power is simultaneous with the power that realizes it.

It seems, then, that intuitions and arguments for the necessity of the past and present dovetail nicely with intuitions and arguments that we can identify possibilities with the realizations of powers. But can we? Are there possibilities which are not the realizations of powers—or realizations of powers which are not possibilities?

Ian Hacking (1975) pointed out forty-five years ago that there are two basic families of modal locutions in English, illustrated by 'possible that' and 'possible for'. He suggested that the former was connected with epistemic modal idioms and the latter with alethic modal idioms. Hacking's suggestion has not been taken up, and it is the 'possible that' locution that is typically used to translate sentences of modal logic, but there is a connection between the 'possible for' locution and the concept of power that is worth considering. While there seems no straightforward way to translate a claim like 'It is possible that Etienne Bacrot will be the next world chess champion' into talk of powers, 'It is possible for Etienne Bacrot to be the next world chess champion' translates easily into 'Etienne Bacrot has the power/capacity to be the next world chess champion'. What now of the relations between the two families? If Etienne Bacrot has the power/capacity to be the next world chess champion, is it possible that he be such? Conversely, if it is possible that he will be the next world chess champion, does it follow that he has the power/capacity to be such?

The short answer to the second question would appear to be 'no'. We need not suppose that every claim of the form 'A can be in state S' can be analysed into a power ascription with A as its subject. Already in the late eleventh century we find Anselm of Canterbury in chapter 12 of *De casu diaboli* (2008, 211–12) arguing that the claim 'A world can exist' is not equivalent to 'A world has a power to exist' but rather has the logical form 'Something can bring about a world'. With this proviso, we have a considerable latitude in ascribing powers and capacities, and a corresponding latitude in accounting for possibilities in terms of them. Might then we be able to account for the entire space of possibilities in terms of powers?

There are at least two complications. First, a powers view of modality privileges the actual. Since powers are properties and 'A is F' entails 'A is', then where nothing is actual there are not any powers. Hence if, for example, it is or was possible that something could come from nothing, that possibility cannot be given a powers analysis. Such considerations have prompted adherents of powers analyses of modality to propose that there are necessary beings with causal powers. They might also be taken as reason to reject the power analysis, one person's modus ponens being another's modus tollens. The second complication arises from the local character of powers. The relation between a powers account and the 'possible for' locution suggests the typical powers ascription is of the form 'It is possible for A to be F', or perhaps 'A has the power to F'. Such locutions have a noun in subject position and predicate F of it. However, sentences of the 'possible that' variety often have rather different and sometimes more complex structures. Consider 'It is possible that there be more humans than there are'. Even if we parse this as 'It is possible for there to be more humans than there are', there is no obvious way to phrase this in the more typical powers fashion. Powers analysts are thus left with a project!

What of the other side of that equation? Is there at least one possibility corresponding to each power? Powers by themselves are not possibilities even if their actualizations or realizations are. Is there, then, for each power at least one way it might be actualized? Jack Spencer (2017) has recently presented counterexamples to what he calls the 'Poss-ability Principle', the claim that for every power there are possible circumstances in which it is realized. Anselm of Canterbury in chapter 3 of *De libertate arbitrii* (2008, 180) had already noted that the presence of an ability was not sufficient for a possibility (his example was of someone with the ability to see a mountain but in circumstances in which there are no mountains) but had not raised the question whether there were powers for which there were no appropriate circumstances. Spencer's examples suggest there are.

3. Scotus and Ockham Again

This brings us back to the dispute between Scotus and Ockham. We can phrase that dispute as one between someone who thinks that there may be powers for which there is no appropriate time of realization (Scotus) and someone who thinks that idea absurd (Ockham). What gives the dispute bite in its context, and ours, is its connection with the ability to do (and the possibility of doing) otherwise. The connection is mediated by what I call Harding's Paradox—that to be free, an agent needs be able to act differently from the way it does in fact act, but there is no relevant time at which it is able to act differently from the way it does act. Suppose Jones reads Proust at time t and we wonder if she could have done otherwise. The times after the choice is made seem irrelevant to whether Jones could have done otherwise at t. What about the times before? Note that even if at every time t Jones could stop reading Proust before t but reads on, that still leaves open the question whether Jones could stop reading at t. What is crucial is whether at t itself Jones is able to stop reading Proust. But at t, Jones *is* reading Proust and so it is too late to do otherwise. She may well be able to stop reading, but she cannot stop reading while she is reading! It would seem then that the actualization of a rational power is incompatible with the power being a rational power! Harding's Paradox thus suggests that if there are powers at t for outcomes at t that are not actual at t, either they must be unrealizable powers, and hence powers for which there are no corresponding possibilities, or the notion of possibility needs be understood so that what is possible need not be possibly actual. What makes it plausible for Scotus to take the second horn of this dilemma, however, is a feature of the broadly Aristotelian conception of power and actuality we have not yet considered.

Besides priority in time, Aristotle (A 1018b9–1019a14, 2:1608–9) distinguishes another way in which a power can be prior to its realization, one which he terms priority in nature. A may be prior in nature to B but simultaneous in time. In particular, a power may be prior in nature to, but simultaneous in time with, its realization. Scholars seem to agree that Aristotle thought there could be no change without time, but medieval thinkers, committed to the creation of the world by an atemporal God, found this problematic, and Aristotle's notion of natural priority provided the material for an alternative picture—one that decoupled possibility, not from change itself, but from change in time.

Can a power, as a power, coexist with its actualization? At first glance this may seem obvious: even though someone lifting a weight may grow tired and be unable to lift it later, she need not, and she may, while lifting the weight at t, retain the power to lift it at t+. Moreover, we may say that trivially while lifting the weight at t she has the power to lift it at t. *Ab esse ad posse valet consequentia*, as a medieval tag had it, and since what is necessary is actual, *Ab necesse ad posse* as well.

There was (and is), however, one very problematic case. In *Metaphysics* Book 9 chapter 2, Aristotle (1046a37–1046b28, 2:1652–3) had distinguished natural powers, which were such that given the power and circumstances the outcome was determined, from rational powers, which could be actualized in different ways in the same circumstances. The will possessed by rational agents (human, angelic, and divine) was taken to be a paradigmatic case of such a power. And it was generally held that, to be free, an agent needed to be able to exercise its will in more than one way, or at least be able to exercise it or not. To accommodate cases like this, and apparently in the light of considerations very like those involved in Harding's Paradox, medieval thinkers borrowed from discussions of Aristotle's (231a2–241b20, 1:390–407) account of the continuum in *Physics* Book 6, the notion of *signa* or, in Scotus's terminology, *instantes* of nature. These were indices which reflected the natural priority relations involved. In particular, thinkers like Henry of Ghent (1520, art.59, q.2) and Scotus (1963) proposed that at a single moment of time there might be several *signa* or *instantes* of nature relative to which claims might be evaluated. What might be true at a given time relative to one of these might be false at that time relative to another. Thus, for example, Henry could reconcile the theory of the Immaculate Conception of Mary, the mother of Jesus, with the claim that every human inherits the stain of the sin of Adam and Eve by positing that at the first moment of her existence Mary had the stain at a prior *signum* but was absolved of it at a posterior one. Thus, there was no time at which she was not immaculate. Scotus in his *Ordinatio* (1963, d.39) applied this idea to the case of the will, arguing that at the time of choice there is a prior *instans* of nature at which the agent has the power to choose a result or refrain from the choice and a posterior *instans* of nature at which the choice is made and the power to do otherwise cannot be exercised.

The analogy between these *signa* or *instantes* and possible worlds or possible situations is compelling and has led scholars to suggest that we find here the origin of the notion that there can be incompatible possibilities (in different possible worlds) at a given time. I think this somewhat oversells the case. Within the Aristotelian picture of powers and their actualizations there is already the idea that at a time there may be incompatible possibilities for a given future time. What Henry, Scotus, and their comrades introduce is the thought that the present is not necessary, and so there may be (even are) powers at a time that are unrealizable at that time.

Does this then uncouple possibility from time? Henry and Scotus do not themselves take this step—they apparently continued to accept the necessity of the past—but it is taken in the next generation by thinkers influenced by them, and although the intuitions undergirding the connection between time and possibility remain strong enough for Descartes to consider the necessity of the past clear and distinct, the full-blooded analysis of modality in terms of quantification that we find in Leibniz is not far away.

What, though, of the connection between powers and possibilities? If we look closely at Scotus's device of *instantes* of nature at a given time, we find there no mention of possible situations. Instead, we find that at time t, relative to a prior *instans* there is the will with its power to act or not act. At time t, relative to a posterior *instans* it acts, and so if we posit it to be not acting at that posterior *instans* of nature, we can derive the contradiction that it acts and does not. Scotus takes it as sufficient for it being possible that the will not act at t that there be some *instans* of nature at t such that if we posit it not acting relative to that *instans*, we get no contradiction. Both *instantes* of nature are actual at t (if *instantes* of nature are actual at all), so there is no introduction here of possible situations or worlds. There is instead an appeal to what would or might happen were certain claims posited in what we might regard as fragments of the actual world.

4. Conclusion

Where does this leave us? Suppose we grant that there is no power, ability, potentiality, or the like the exercise of which now would bring about an alternative to what is now present. Does it follow that there is no sense to be made of such an alternative? Even if such strong metaphysical principles as the Necessity of Origin force on us the thought that nothing that is actual is part of such an alternative, might there still not be one entirely alien?

I see no way to deny this as long as one thinks that something can come to be from nothing. If things might just happen even though there are no abilities, capacities, powers, propensities, objective chances, tendencies, or the like which account for their happening, then there is no obvious bar to there being right now (but not actually) 'alternatives' not only to the present but to the past as well. They would not, however, be alternatives for us and the other things that actually are but, at best, situations in which our doppelgängers (or counterparts) interacted with doppelgängers (or counterparts) of the things around us. The space of possibilities may extend beyond that which can be analysed in terms of powers and capacities but, it would seem, at the price of these not being *actual* possibilities. Suppose, then, that nothing can come from nothing, that the origins of things are essential to them, that things now are necessarily what they have become, and that the mere existence of a power as contrasted with its realizability is not sufficient for a possibility. What then?

Could I have chosen to write these lines at the Siam Pung Thai coffee house instead of Betty's in West End? Of course! Could a person just like me except that he did write lines just like these at a place just like the Siam Pung *be me*. Certainly not. Suppose I will write next time at the Siam Pung but can at Betty's (not likely but surely possible). Does my continued identity depend on that choice? Barring catastrophe, surely not. Each of us can do many things we will not do, but nothing which did what we did not do can now be us. That is the modal difference time makes. As Oliver Heaviside is reported (by Mark Wilson) to have said, 'Logic is Eternal, it can wait'. Possibilities, on the other hand, often have to be seized or foregone. *Carpe diem*!

References

Anselm of Canterbury. 2008. *Major Works*. Ed. Brian Davis. Oxford: Oxford University Press.

Duns Scotus, John. 1963. *Ordinatio, Liber I*. In volume 6 of *Opera omnia*. Ed. Pacifico Perantoni, Carolus Balic, Barnaba Hechich, and Josip Percan. Vatican City: Typis Polyglottis Vaticanis.

Ghent, Henry of. 1520. *Summa quaestionum ordinariarum*. Paris.

Grünbaum, Adolf. 1963. *Philosophical Problems of Space and Time*. New York: Alfred A. Knopf.

Hacking, Ian. 1975. *The Emergence of Probability*. Cambridge: Cambridge University Press.

Harré, Rom, and E. H. Madden. 1975. *Causal Powers: A Theory of Natural Necessity*. Oxford: Basil Blackwell.

Hintikka, Jaakko. 1973. *Time and Modality*. Oxford: Oxford University Press.

Knuuttila, Simo. 1993. *Modalities in Medieval Philosophy*. New York: Routledge.

Kripke, Saul. 1980. *Naming and Necessity*. Cambridge: Harvard University Press.

Normore, Calvin. 2003. 'Duns Scotus's Modal Theory'. In *The Cambridge Companion to Duns Scotus*. Ed. Thomas Williams, 129–60. Cambridge: Cambridge University Press.

Ockham, William of. 1978. *Tractatus de praedestinatione et de praescientia Dei*. In vol. 2 of *Opera philosophica et theologica*. Ed. the Franciscan Institute (St. Bonaventure University). St. Bonaventure, NY: Editiones Instituti Franciscani Universitatis S. Bonaventurae.

Reichenbach, Hans. 1958. *The Philosophy of Space and Time*. New York: Dover Publications.

Rescher, Nicholas. 1963. *Studies in the History of Arabic Logic*. Pittsburgh: University of Pittsburgh Press.

Shoemaker, Sydney. 1969. 'Time Without Change'. *Journal of Philosophy* 66, no. 12 (June): 363–81.

Spencer, Jack. 2017. 'Able to Do the Impossible'. *Mind* 126, no. 502 (April): 466–97.

Van Fraassen, Bas. 1966. 'Foundation of the Causal Theory of Time'. PhD diss., University of Pittsburgh.

3

Aristotelian Powers, Mechanism, and Final Causes in the Late Middle Ages

Henrik Lagerlund

Domandatur causam et rationem quare

Opium facit dormire.

A quoi respondeo,

Quia est in eo

Virtus dormitiva,

Cujus eat natura

Sensus assoupire.

You ask for the cause and the reason why

opium makes one sleepy.

To which I respond,

That it is because it has

A dormative power,

By which nature it is

To make the senses doze off.

Molière, *Le malade imaginaire*

In his famous play *Le malade imaginaire*, Molière includes an interlude about a teacher interrogating a student about why opium puts people to sleep. The right answer is supposed to be because opium has a dormative power. In 1673, when the play premiered, this was supposed to be satirical and funny, as well as an example of the many absurd things scholastic medicine and natural philosophy had claimed. Such things as powers, the new science of the seventeenth century had moved well past, and they now belonged on the pile of silly, worthless ideas of the past, or at least that was what they thought at the time.

Other examples from this time ridiculing the notion of causal powers have a lot to do with what was taken to be the anthropomorphizing of nature,

Henrik Lagerlund, *Aristotelian Powers, Mechanism, and Final Causes in the Late Middle Ages* In: *Reconsidering Causal Powers: Historical and Conceptual Perspectives*. Edited by: Benjamin Hill, Henrik Lagerlund, and Stathis Psillos, Oxford University Press (2021). © Henrik Lagerlund.
DOI: 10.1093/oso/9780198869528.003.0004

which explains events by attributing characteristic human properties to them, like attributing appetitive powers to keys. The key has the power, capacity, or ability to open the door. In the general framework of mechanism this was not seen as an explanation. But what is exactly wrong with this sort of explanation? Well, they seemed to think that nothing had been explained when it is said that the opium smoker feels sleepy because of the opium's dormative power, that is, it is just restating what needed to be explained in the first place, they thought, preferably by spelling out the mechanism behind it.

To further get a sense of what the objection actually comes to, we can compare it with ordinary folk explanations, like when we say that these pills make me drowsy (like my allergy medication, for example). We have a tendency in ordinary language to say things like 'these pills have the power or ability to make me drowsy'. I say, 'It is dangerous to take these pills before driving.' 'Why?' 'They make me drowsy.' 'Why?' 'Because they have that effect on me, or that power on me!' The why-question used here is normally associated with final causality and not efficient causality. Powers and final causality are at least historically connected. To see that let me turn to Aristotle.

1. Aristotle and Powers as Goal-directed

The Greek word used by Aristotle to talk about powers or capacities is 'dunamis [δύναμις]'. His definition of 'dunamis' in *Metaphysics* Book 5 is that it is a source of movement or change. He writes:

> We call a power [*dunamis*] a source of movement or change, which is in another thing or in the same thing qua other, e.g. the art of building is a power which is not in the thing built, while the art of healing, which is a power, might be in the man healed, but not in him qua healed. Power then is the source, in general, of change or movement in another thing or in the same thing qua other, and also the source of a thing's being moved by another thing or by itself qua other.   (A 1019a15–19, 2:1609)

The first and most basic use of *dunamis* is as an agent power, or as a causal power to implement a change in something other than itself, like the ability to heat or to see. Aristotle uses *dunamis* in this way for both natural phenomena and for human or animal activity (powers or abilities of the soul). As such *dunamis* occurs in a number of different works, from biology and physics to ethics. In *Metaphysics* Book 9, he makes the idea of *dunamis* more complicated and fleshes out the concept in quite some detail.

The intuitive idea is that things have abilities to do or to be other than they are. Substances living or non-living have potentialities to do or be other than they are. It is a very basic intuition for Aristotle. In *Metaphysics* Book 9, he brings together several important elements that all have over the centuries come to define Aristotelianism, like activity, potentiality, possibility, power, capacity, ability, substance, quality, cause, etc.

I would like to stay with the idea of final causality, since it emphasizes in what way one could confuse powers with mental abilities and brings out why this became such a common argument against powers. The idea of an end or a goal for the activity is essential to the Aristotelian notion of *dunamis*. Aristotle (1050a9–17, 2:1658) says in *Metaphysics* Book 9 that 'the actuality is the end, and it is for the sake of this that the potentiality is acquired...animals do not see in order that they may have sight, but they have sight so that they may see...matter exists in a potential state, just because it may come to its form; and when it exists actually, then it is in its form.' It always moves from something to something and, without the goal or the end, the movement would be accidental or haphazard. The power needs to 'know' where it is going, so to say. To a modern ear this undoubtedly sounds like there is something like a desire or even 'cognition/knowledge' built into the substances themselves making them go downwards, that is, something which inheres in solid bodies in the case of gravity. It makes the power look like 'little souls' or 'minds'. Descartes puts it well in the Sixth Replies to the *Meditations*. He writes: 'I conceived of gravity as if it was some sort of real quality, which inhered in solid bodies; and although I called it a "quality", thereby referring it to the bodies in which it inhered, by adding that it was "real" I was in fact thinking that it was a substance.... But what makes it especially clear that my idea of gravity was taken largely from the idea I had of the mind is the fact that I thought that gravity carried bodies towards the centre of the earth as if it had some knowledge of the centre within itself" (AT 8a:441–2; CMS 2:297–8). Obviously, bodies cannot have such powers, Descartes goes on to argue. Whatever powers they have they must reduce to extension and the movements of extended bits of matter. However, the idea that these are powers or qualities that normally are attributed to minds was certainly lost on Aristotle, who had a more uniform notion of nature than Descartes. The idea of natural powers as close to or identical to mental powers was, however, emphasized just before Descartes in the works of Francisco Suárez (*DM*).

When Suárez sets out to defend substantial forms, which is the principle of causal powers for him, he starts with the human soul, which for him is the most obvious example. He writes: 'The first argument for the existence of substantial forms is that a human consists of a substantial form as an intrinsic cause; therefore so does all natural things.' And a little further on he adds: 'this soul is a substantial form, as we will show below, that name "substantial form" signifies nothing other than a certain partial substance which can be united to matter in such a way that it composes with it a substance that is whole and per se one, of which kind is a human being' (*DM* 15.1.6, 1:499a–b; Suárez 2000, 20).

The whole argument following this is set up so that all natural substances are explained in relation or analogous to the human soul. Obviously, they get less and less sophisticated and more material, but they retain a sense of the substantial form of human beings. Given the strong analogy of natural, material substance with the human substance, it is very easy to get the impression of the substantial forms attached to matter as 'little souls', and it seems clear that Descartes's understanding of these qualities as quasi-substances is very close to Suárez's. In fact, reading Suárez next to Descartes's Sixth Reply, it seems obvious that Descartes has Suárez in mind, or his view in mind, when writing (for more see Hattab 2009).

Even though the association of powers with 'little souls' becomes important in the seventeenth century and is, it seems to me, actually one of the better arguments against Aristotelian powers from this time, it does injustice to the Aristotelian tradition of the Middle Ages, however, since that tradition is very diverse and contains many different views. Suárez's view was also rather idiosyncratic. There is, for example, a clear tradition starting in the fourteenth century with William Ockham and progressing well into the sixteenth century even up until the time of Suárez that shows many similarities with the view held by the mechanical philosophers of the seventeenth century.

2. A Medieval Mechanism

Mechanism as we have come to know it rests on an assumption, which came to be enormously influential and still today underlies our modern scientific world view, that has to do with how wholes are related to their parts. Both material bodies and events in nature can be analysed and divided up into their most elementary parts, which determine the whole. The whole must be understood from its parts and it is also ultimately reduced to its parts. Hence, it rests on an assumption about mereology.

This way of thinking about nature is also codified in the discussion of method. To divide something into its parts is called analysis, and the construction of the whole from parts is called synthesis. Both Descartes (AT 5:19; AT 10:379; CSM 1:120; CSM 1:20) and Galileo famously give descriptions of the two methods. Galileo describes what he calls a *metodo risolutivo* and a *metodo compositivo*, and, as we now know (see Randall 1961), his method derives from the fourteenth century and finds its way to Galileo through the Padua Aristotelian tradition.

Perhaps it comes as no surprise, then, that thinking about material bodies or material substances in a mereological way also derives from the fourteenth century. The very specific view of nature as material, extended parts constituting the whole body or substance can be traced back to nominalists like Ockham and Buridan (Normore 2006; Lagerlund 2012). Ockham's ontology includes individual substantial forms, individual accidental forms, and individual matters. A composite substance is composed of its essential parts, namely, its matter and form. Besides its essential parts, a substance also has integral parts, like flesh, bones, hands, and so on. Anything with integral parts is extended and material. All things with essential parts are composed of matter and, hence, also have integral parts and are extended, according to Ockham. Everything extended is a quantity, and every quantity is divisible into quanta. Hence, there are no indivisible matters and substances.

Ockham insists that a substance is nothing but the parts that make it up. Each part of a substance is actual and not dependent on anything to make it actual, he argues. Also, all the forms in all material substances are extended. In a piece of wood, all the forms are extended just as the matter they inform. The only non-extended forms are the human intellectual soul, angels, and God on Ockham's view. Angels and God are, however, outside nature, and hence the only non-extended or immaterial form in nature is the human soul.

As all material substances change, and since change in Ockham's view means that some part is replaced, the problem of unity or identity over time and change becomes particularly problematic. Given this view of substance, it is not clear what constitutes the unity or identity over time of a substance. For him, a substance is a substance due to its parts, and all parts are individual parts of the substance. No substance on this view is, after a process of growth, for example, numerically the same as before. Ockham does not explicitly address this problem, but Buridan does (Normore 2006; Lagerlund 2012).

To determine the issue, Buridan outlines three ways in which something can be numerically the same over time, namely (1) totally, (2) partially, or (3) successively. Something that never gains or loses a part is totally the same over time. Something is numerically the same in its entirety, as he puts it, if its integrity is completely preserved. On this view only indivisible substances are totally the same over time. There are only three such things, according to both Ockham and Buridan, as we have seen, namely God, angels, and the human soul. Things that are partially the same over time are such things that have a principal part that is totally the same over time. In nature, it is only humans that are partially the same over time since our principal part, the intellective soul, remains totally the same. All other substances have a weaker form of sameness. Buridan never explicitly tells us what is required for something to be successively the same over time. He (Buridan 2016, 10) gives an example of the River Seine that is said to be the same river over a millennium because the water parts succeed each other continuously. This is not sameness, properly speaking, Buridan seems to think, since there is nothing of the whole that is the same over time, but rather there is a succession of entities, related enough so that the same name can be applied.

Although no composite substance in nature, except humans, remains the same after growth or decay, one can say on Buridan's view that a horse or a river is the same over time because of the continuity of its parts, and this sort of sameness does not require that any given part remain through the change. On this view, then, nothing except a human in nature has an identity or unity stronger than a that of a heap or pile of things or a river, which all have the same unity or identity. An animal, for example, is the same over time in the same way as the River Seine is the same over time. From birth to death, the animal is the same because there is a succession of parts succeeding each other occupying the same spatio-temporal location. This is true of a heap as well, and there is no other unity to an animal. This succession of entities over time allows for the appearance that the name picks out the same entity over time, although it in reality picks out several entities continuously succeeding each other. There is a lot of discussion of this view of material substance in commentaries on Aristotle's *Physics* after Buridan all the way up to the sixteenth century. They even introduced a question dealing with whether a whole substance is its parts. It was, hence, a well-developed view of material substance by the time of Suárez (for a comparison of this view with seventeenth-century views see Lagerlund 2012).

3. Powers and Final Causality in Ockham and Buridan

On this view of substance there is no room for powers or final causality. In general, Ockham wanted to reduce his ontological commitments, and as part of this program he eliminated powers from among the kinds of things exist-ing in the world. Starting from his predecessor John Duns Scotus's distinction between natural and rational powers, Ockham claimed that there are no powers in nature that arise from nature itself. All power is bestowed on it by God or some other active power like the human mind. Ockham, like Descartes, was a dualist with respect to mind and body. He attributed power or activity to minds alone. The mind has the 'rational' power to will.

Final causality is, as noted, fundamental for Aristotelian physics and for the notion of a causal power. The critique of final causality was therefore a corner-stone of the so-called mechanical philosophers' attack on Aristotelianism. The final cause or end for which natural agents were supposed to act came in for particularly harsh criticism: as seen above, the Aristotelians, in attributing ends to natural changes, were said to have confusedly bestowed minds on natural agents. In this respect powers and ends in nature were similar. Spinoza, for example, argues in the appendix to *Ethics I* that final causality turns nature upside down and puts the effect before the cause. Suppose, for example, that a stone, in falling off a building, kills someone. According to Spinoza, the Aristotelian must hold that the outcome—death—is the *end* toward which the efficient cause—the stone—acts when it falls. The outcome is then held to be the cause, and the stone's action, or the stone itself, an effect. This is what was meant by turning nature upside down and only minds could act for an end. Descartes, for his part, argued that since we cannot know God's intentions natural philosophy has no business looking for ends in nature. Instead, the natural philosopher should merely try to discover the laws gov-erning motion. There is a close connection between the refection of final causality and the rise of laws of nature, which we cannot consider here, how-ever, but there is also, as seen, a close connection between causal powers and final causality and both are associated with minds in the fourteenth century, certainly by Ockham and Buridan.

As I (Lagerlund 2011) have claimed previously, Ockham and Buridan had already in the fourteenth century fundamentally reshaped the discussion and it was their debate that affected future generations, as, for example, can be seen in Suárez's treatment of final causality (Schmaltz 2008; Åkerlund 2011). Ockham (1991) made it clear in his *Quodlibetal Questions* IV, q. 1, what he thinks final causality is: 'nothing other than its being loved and desired efficaciously by an agent, so that the effect is brought about because of the thing that is loved.' The object of someone's love is the end of that person's actions, which are caused efficaciously for the sake of that end. The efficient cause of someone's action is the will, but the final cause is the object loved. He is very clear further on in the same *Quodlibetal Question* that there is no final causality in nature and draws a sharp distinction between free agents and natural agents. The final cause is in the mind and not in nature on this view.

Buridan again works out the details of Ockham's suggestion. His rethinking of final causality is motivated by the very problem that Spinoza would highlight 300 years later. Buridan argues that every cause is naturally prior to the caused thing, but the end is not naturally prior (for discussion see Lagerlund 2011). The problem is thus how something that is posterior to its effect can be a cause. To do so, it must act backwards in time, since causes are naturally thought about as that which brings something about. The argument that worries him most seemed, however, to be the following. Take the traditional Aristotelian picture that the cause of the doctor's prescription of a certain medicine is the health of the patient. By analogy, then, it must follow that the cause of God's action of creation is the world being created. But this implies that things inferior to God are causes of His actions, which is clearly absurd. Buridan takes this to be a knock-down argument of final causality.

On Buridan's view the final cause is instead internal to us as our desire or reason for acting. He hence reduces all final causality both in nature and in mind to efficient causality. On such a view the argument that seems to have worried him the most, namely the argument that the world is the cause of God creating it, is not a problem. The world is not the cause of anything, since the world only exists as an intentional object in the mind of God and as such it is an efficient cause of God's act of creation (for discussion see Lagerlund 2011).

In the mid-fourteenth century, we, hence, see the development of a view of nature very similar to the prevailing one associated with the seventeenth-century mechanical tradition. We see a sharp distinction being drawn between minds (intellectual souls) and nature—minds do not belong to nature according to Ockham and Buridan. Final causality is eliminated from nature, and in Buridan from minds as well.

4. Buridan on Powers as Dispositions

Unlike Ockham, Buridan does perhaps not fully eliminate causal powers, however. He develops a notion of causal powers in terms of dispositions, but it is ultimately unclear what he means by a disposition (see Zupko 2018 and Klima 2018 for different attempts to narrow down what Buridan might mean by 'habit' and 'disposition'). Buridan sees all form and all change as dispositions. He writes: 'I state here that every natural change requires an agent that acts and produces some form or disposition. Natural changes also require a subject which receives that form, whether that form is a quality, or quantity, or the motion that we call "locomotion," or even a substantial form. This [produced] form or disposition is the change itself' (Buridan 2001, 3.6.1). In his *Physics*, he also defines motion as a disposition. He writes:

> Motion is the act of a being in act, because it is the act of that which moves and of that which is moved. But these verbs 'to move' and 'to be moved' signify to be in act according to a disposition (*dispositionem*) just as 'to be able to move' and 'to be able to be moved' signify to be potentially. And universally, by the definition of the meaning of the term it is clear that all act by which something is said to be in act, when before it would be said to be in potency, is the act of a being in act.   (2016, 3.10)

Objects have dispositions, according to this view, namely, either active or passive dispositions. An active disposition is for an object to move or change in some way, while a passive disposition is for an object to be able to move or change, although it is currently not doing so.

'Disposition' is a rather broad and inclusive term for Buridan. It covers, of course, habits, as can be seen from the following passage: 'Nevertheless, Aristotle in the *Categories* remarks that although the term "disposition," taken strictly, is distinguished from the name "habit," still, sometimes the name "disposition" taken broadly, insofar as it extends to all the aforementioned qualities in account of which things are well- or ill-disposed to operate, whether these qualities are removable easily or with difficulty; this is why he says that "disposition" is a broader term than "habit" and that every habit is a disposition, but not every disposition is a habit' (2001, 3.5.2). All mental capacities are also disposition: 'These [two kinds of] qualities agree in that both are dispositions to act or suffer well or unwell. But they differ in that if they are innate, deriving from the strong and weak principles of generation, then they are called "natural capacities" and are said to belong to this species, but if they are acquired by learning or custom or in some other way, then they are said to belong to the first species' (3.5.3). What Aristotle terms 'power [*dunamis*]' is for Buridan a disposition. Even though he develops a notion of body or corporeal substance as extended parts arraigned in a certain way, he does not seem to think that it is enough to explain why these bodies do what they do. Buridan therefore attributes dispositions to these bodies that are not simply reducible to the parts that make up that body. In the *Metaphysics* he says the following:

> It is true that Aristotle would not have conceded that matter and form were united and composed together through some disposition added to the subject like whiteness and human, because he thought substantial as well as accidental form inseparable from their subjects through any power except for the corruption of the subject, but because we hold from faith that they can be separated and conserved separated, therefore it seems necessary to hold and added disposition or inherence, namely an inherence of the form in the matter.   (Buridan 1964, 7.2, quoted in Normore 2017, 68 n11.)

Even though he here is talking about the relation between the human soul and the body, it suggests he is happy to think about dispositions as something over and above a part, but a property added to it. Is this not just another form? And is this not adding back into his metaphysics something that his conception of body and material substance, outlined above, was meant to reduce away? Perhaps, but I do not think any heavier metaphysics should be read into this. '*Dispositio*' in classical Latin means, foremost, an arrangement or an order. It is, hence, possible to interpret Buridan as using the word to mean something similar. A disposition is, then, not at all a power in the Aristotelian sense but a mere arrangement of the parts of a substance. Every change or motion would, hence, simply mean a new arrangement of the parts. However, if this notion of disposition is supposed to be understood, it must be, it seems to me, in a very thin sense. So, even though Buridan uses dispositions to talk about all the things traditionally attributed to Aristotelian powers, he has thinned out the notion so much that it retains nothing of its original metaphysics and certainly it has lost any connection with final causality. He, like Ockham before him, ends up with a view of nature utterly divorced of any so-called mental powers and a view in line with that of many mechanical philosophers of the seventeenth century.

5. Conclusion

Ockham eliminated all powers and all final causality from material substances, and associated them with minds, or immaterial substances. Buridan eliminated final causality from minds as well, but he retained a notion of power in terms of dispositions, which he attributed to material substances as well. In his defence of substantial form, in the sixteenth century, Suárez, I think, can be seen as reintroducing a much stronger notion of active powers to nature, which at least seems to bring with it final causality as well, by modelling all substantial forms on the human mind. It adds to a material substance all the things that Ockham and Buridan had pushed into minds. It is this view that Descartes and other mechanical philosophers are reacting to in the seventeenth century.

References

Åkerlund, Erik. 2011. ' "Nisi temere agat": Francisco Suárez on Final Causes and Final Causation'. PhD diss., Uppsala University.

Buridan, John. 1964 (1588). In Metaphysicam Aristotelis quaestiones. Paris. Repr. Frankfurt: Minerva.

Buridan, John. 2001. Summulae de dialectica. Trans. Gyula Klima. New Heaven: Yale University Press.

Buridan, John. 2016. Quæstiones super octo libros Physicorum Aristotelis (secundum ultimam lecturam) libri III–IV. Ed. Michiel Streijger and Paul J. J. M. Bakker. Leiden: Brill.

Faucher, Nicolas, and Magali Roques, eds. 2018. The Ontology, Psychology, and Axiology of Habits [Habitus] in Medieval Philosophy. Dordrecht: Springer.

Hattab, Helen. 2009. Descartes on Forms and Mechanisms. Cambridge: Cambridge University Press.

Klima, Gyula. 2018. 'The Metaphysics of Habits in Buridan'. In. Faucher and Roques 2018, 321–31.

Lagerlund, Henrik. 2011. 'The Unity of Efficient and Final Causality: The Mind/Body Problem Reconsidered'. British Journal for the History of Philosophy 19 (4): 585–601.

Lagerlund, Henrik. 2012. 'Material Substance'. In The Oxford Handbook of Medieval Philosophy, 468–85. Ed. John Marenbon. Oxford: Oxford University Press.

Normore, Calvin. 2006. 'Ockham's Metaphysics of Parts'. In 'Parts and Wholes', ed. Wolfgang Mann and Achille C. Varzi, special issue, Journal of Philosophy 103, no. 12 (December): 737–54.

Normore, Calvin. 2017. 'Buridan on the Metaphysics of the Soul'. In Questions of the Soul by John Buridan and Others, ed. Gyula Klima, 63–75. Dordrecht: Springer.

Ockham, William. 1991. *Quodlibetal Questions.* Trans. Alfred J. Freddoso and Francis E. Kelley. 2 vols. New Haven: Yale University Press.

Randall, John. 1961. *The School of Padua and the Emergence of Modern Science.* Padua: Editrice Antenore.

Schmaltz, Tad. 2008. *Descartes on Causation.* Oxford: Oxford University Press.

Suárez, Francisco. 2000. *On the Formal Cause of Substance.* Trans. John Kronen and Jeremiah Reedy. Milwaukee: Marquette University Press.

Zupko, Jack. 2018. 'Act and Disposition in John Buridan's Facutly Psychology'. In Faucher and Roques 2018, 333–46.

4

Agency, Force, and Inertia in Descartes and Hobbes

Deborah Brown

One body collides with another, causing it to move. A natural supposition is that the change in the second body is produced by the action of the first. Since its earliest appearance, Descartes's physics has been thought to make the idea that one body exerts an active force (*vis*) unintelligible. For all its oversimplifications, Aristotelian physics, by contrast, had little difficulty accommodating the intuition that bodies have intrinsic causal powers, upon which distinctions between action and passion, agent and patient, are grounded. For Aristotle, each thing or kind of thing possesses a definitive set of intrinsic *potentiae*—powers to produce change and potentialities to undergo change—that are conceptually tied to the notion of its substantial form. The collective exercise of such powers accounts for all observable natural change. Regarding such *potentiae* as occult qualities,[1] more a name for a problem than a solution, mechanists like Descartes and Hobbes sought instead to explain change and differentiation in terms of the law-governed interactions of matter in motion. Central to this picture is the conception of bodies as essentially *inertial*. Stripped of intentions and natural ends or final causes, bodies do not change unless interfered with by something larger in magnitude or bearing a greater quantity of motion. But if bodies lack active causal powers, the question naturally arises what *is* responsible for all the variation and change that we observe in nature?

The debates about causation that dominate the seventeenth century—debates about whether, in addition to God's efficient causality, there are real,

---

[1] The pejorative sense of 'occult' stems from the early modern period where it becomes associated with mysticism and pseudosciences such as alchemy and astrology. Prior to the 1630s, 'occulte' (Middle French) or 'occultus' (Latin) from the verb 'occulere' (to cover over or hide) would have signified only that the phenomenon is concealed, hidden, or secret. See, for example, https://www.etymonline.com/word/occult.

Deborah Brown, *Agency, Force, and Inertia in Descartes and Hobbes* In: *Reconsidering Causal Powers: Historical and Conceptual Perspectives*. Edited by: Benjamin Hill, Henrik Lagerlund, and Stathis Psillos, Oxford University Press (2021).
© Deborah Brown.
DOI: 10.1093/oso/9780198869528.003.0005

secondary causes in nature; debates about whether bodies, especially organic bodies, move for the sake of some end (final causation); and debates about whether and in what sense incorporeal substances act upon corporeal substances (and vice versa)—all hinge on this question of what accounts for the apparent activity in nature. To those who disputed the idea that bodies possess intrinsic causal powers, there appeared an inevitable slide from the inertial conception of matter to the conclusion that the power to move bodies belongs either to God and God alone (occasionalism[2]) or to God and his incorporeal *instruments* (vitalism). These instruments of God were variously construed as vital elements (Harvey (1628) 1910, 68)[3], the Knowing Principle (More 1925, 39–40), plastick natures (Cudworth (1678) 1837-8, 221–4), or monads (Cudworth (1678) 1837-8, 1:242–3; Leibniz (1714) 1898). For many of Descartes's critics, the exhaustive claims to adequacy of his 'geometrical physics' (AT 8A:78; CSM 1:247) lend themselves to a static conflation of body with space, with too few conceptual resources to explain truly dynamic phenomena (Gabbey 1971/2; Grosholz 1988). Against this background, Descartes's occasional references to mysterious 'forces' of bodies appear as much to rely on occult qualities as the scholastics' appeal to active and passive powers.

I want to claim neither that Descartes is clear about what he means by 'force' nor that he adequately accounts for it within the scope of his geometrical physics. I also do not want to claim that Descartes was always faithful to the principles of his own mechanistic philosophy, which, as Emily Grosholz describes it, relies all too much on 'loose analogy and a quite imaginative array of "mechanisms"' (Grosholz 1988, 238). But none of the available interpretations of 'force' in Descartes's physics—be they reductionist, in the sense of equating force with some attribute or mode of extension,[4] or be they, in the

---

[2] Prominent occasionalists in the period were Nicolas Malebranche, Louis de La Forge, and Géraud de Cordemoy. For Malebranche, the conclusion that God is the only cause follows because it is only in God that we perceive a necessary connection between action and effect: 'As I understand it, a true cause is one in which the mind perceives a necessary connection between the cause and its effect. Now it is only in an infinitely perfect being that one perceives a necessary connection between its will and its effects. Thus God is the only true cause and only He has the power to move bodies' (*Search* 6.2.3, 450). There are echoes in Malebranche's necessitarianism of the tradition of per se causality (e.g. Aristotle, *Physics* II), which holds that, for a causal relation to be intelligible, there must be a necessary (i.e. non-accidental) and conceptual connection between the cause and the effect. For a recent defence of the occasionalist reading of Descartes, see Ott 2009, 64–77.

[3] William Harvey posited an intrinsic contractile power to the heart.

[4] See, for example, Westfall (1971). Reductionist readings are grounded in statements by Descartes like the following: 'The only principles which I accept, or require, in physics are those of geometry and pure mathematics; these principles explain all natural phenomena, and enable us to provide quite certain demonstrations regarding them' (AT 8A:78; CSM 1:247).

wake of failed reductions, attempts to identify force with God's power,[5] or be they mixed theories that attribute force to bodies and God,[6] or be they deflationary in regarding attributions of 'force' to bodies as simply loose talk[7]— captures a fundamental fact about the way forces are described. The burden of this chapter is to draw (or redraw) attention to the fact that, when describing motive force, Descartes is apt to use the terminology of 'tending' or 'tendencies' (from the verb *tendere*) and to treat these tendencies as real and causally explanatory. I say 'redraw' because here I am taking the lead from T. S. Champlin (1990/1, 120–1), who traces a line from Descartes through J. S. Mill and W. Stanley Jevons to Peter Geach, and, more recently, to Nancy Cartwright, in thinking of laws as tendency laws.[8] These tendencies, as we shall see, are reducible neither to extension nor to its modes and attributes. Indeed, attempts to reduce these tendencies (for example, Hobbes's) generated more puzzles than they solved. And the suggestion that such tendencies are really tendencies of God will, I hope, strike the reader as ludicrous.

Tendency talk is not, I submit, a mere *façon de parler* on Descartes's part. As will become clear, two points are salient about his use of the terminology. First, tendencies presuppose but are not reducible to motion or any other mode or attribute of extension. In modern parlance, we could say that tendencies *supervene* upon but are *not reducible* to motion. Perhaps even that will turn out to be too strong a claim—perhaps all that is required is that tendencies presuppose the *possibility* of motion. But let's not prejudge the

---

[5] Hatfield (1979, 113) notes that the imparting of motion is not something that can be explained in terms of 'matter (extended substance) in motion (where this motion is explained kinematically)'. Since it is God's immutability that is responsible for conserving motion through its transfer from one body to the next, it is God who imparts the force by which changes of state occur (129). Cf. Garber (1992, 294), who points to texts where Descartes explicitly attributes force to bodies.

[6] For mixed views grounding force in extension and God, see Gueroult 1980, Gabbey 1971/2, and Des Chene 1996. Des Chene (341), however, notes the tension in mixed theories in that 'Descartes...defines matter as *res extensa* in part just to exclude active powers. Indeed, in his physics, if not his psychology, the very notion of *concurrence* begins to lose its grip. Concurrence is *co-action*, not action *simpliciter*. But if bodies do not act, there is no concurrence; there is only the outright effecting of change in the physical world by God alone.' Des Chene nonetheless thinks that sense can be made of Descartes's use of 'force'. Predicated of bodies, 'force' is a way of characterizing God's action in terms of its effects on bodies. Force is real, but the force attributed to bodies just is divine action construed in terms of its effect on bodies. One difficulty with such approaches lies in needing to reconcile how force can be both the passive effect of God's activity on a body and, at the same time, the active power of that body. Another relates to how, if forces are grounded in the unchanging attributes of extension and duration in a body (a reflection of the immutability of their creator), they can vary according to differences in circumstance or over time. See Garber's (1992, 296) observations on this difficulty.

[7] Garber (1992, 298) writes: 'forces...can be regarded simply as ways of talking about how God acts, resulting in the lawlike behaviour of bodies; force for proceeding and force of resisting are ways of talking about how, on the impact-contest model, God balances the persistence of the state of one body with that of another.'

[8] Cartwright (1989, 226) sees 'capacities', a subset of tendencies that define specifically the causal powers of things, as more fundamental than laws.

situation just yet. What is clear is that Descartes distinguishes between force and motion while acknowledging an intimate connection between the two. Second, there is nothing particularly occult about a tendency. The tendency of my car to pull to the left is not the same thing as my car's pulling to the left on actual occasions, but neither is it a 'real quality' in Descartes's pejorative sense—a quality separable from such instances of the car's moving to the left or from the car itself. Thinking about tendencies might just help, therefore, to demystify Descartes's references to active forces.

We must, however, proceed with caution. It would be natural to baulk at the significance of Descartes's tendency talk, at the idea that tendencies are 'real'[9], at the idea that they are irreducible, and at the idea that there could be any unequivocal interpretation of 'force' that includes within its extension the causal powers of both corporeal and incorporeal substances. All this will, of course, need to be teased out as we proceed. But I remain cautiously optimistic, not just at the prospect of clarifying what for Descartes active forces are, but also for gaining some insight into what the law of inertia actually entails. The way in which inertia was thought to exclude the possibility of active forces in Cartesian physics is nicely encapsulated in Louis de La Forge's *Traité de l'esprit de l'homme*. Let us turn then immediately to that text.

1. La Forge on Force and Inertia

The slide from the law of inertia to the conclusion that motive force does not belong to bodies themselves is evident in the following passage from the *Traité de l'esprit de l'homme* by Descartes's occasionalist follower, Louis de La Forge:

> Now if the force which moves is distinct from the thing which is moved and if bodies alone can be moved, it follows clearly that no body can have the power of self-movement in itself. For if that were the case this force would not be distinct from the body, because no attribute or property is distinct from the thing to which it belongs. If a body cannot move itself, it is obvious

---

[9] One could object, for example, that a tendency is nothing more than a way of describing the conditional probabilities of various outcomes given the possibility of certain conditions obtaining. Gibson (1983, 306–7) argues that tendency statements are 'partial truths', conditional statements in which the antecedent conditions can only be partially specified given the impracticality of spelling out all the *ceteris paribus* clauses involved in determining the conditions under which the tendency will be realized. This is in opposition to views that suppose that tendency statements are true and that real tendencies are truthmakers.

in my opinion that it cannot move another body. Therefore, every body which is in motion must be pushed by something which is not itself a body and which is completely distinct from it.   (*THM* 145)

La Forge's argument begins with the supposition that motive force is really distinct from the body moved, an assumption he tries to defend first by observing that if force were not really distinct from the body moved but were instead a mode, it could not be transferred in collisions, and so it could not account for how it is that one body causes a change in another. This is because no mode, having being only as a modification or *way* a substance is, could either separate or be transferred from one substance to another.

Second, if the motive force were not really distinct from the body, it would of necessity 'include in its concept the idea of extension', but clearly it does not. La Forge does not initially provide an argument in support of this assertion but later offers a thought experiment designed to convince us that force cannot be explained through the concept of extension. Imagine, he says, that God removes all motion from the universe. Since all distinctions between bodies depend upon motion, the result would be a 'formless mass'. Could anything in this indefinitely extended plenum initiate motion? No, and indeed such a scenario invites the following dilemma: for motion to commence, either a single part or the whole of extension must be capable of initiating motion. There are no other alternatives. But no part could move itself without having to move the entire plenum, which is absurd. And what, moreover, would determine the *direction* in which that part moves? But neither could the whole plenum move itself, for if God had created the universe entirely at rest and God is immutable, what power in this formless mass could put asunder what God has created? La Forge declares that it is easy to answer the question of whether bodies have the force to move themselves in the negative '*because extension, in which the nature of body in general consists and which is the only quality which it retains in this condition, is not active*' (*THM* 146). He then proceeds to reason, contrary to popular opinion, that the power to move bodies belongs to spiritual substances alone.

It may be objected that by his thought experiment La Forge has proven too much. For if from the fact that extension can exist without motion it follows that motive force and extension are really distinct, should we not also conclude that motion too is really distinct from extension and 'does not include in its concept the idea of extension'? Such a suggestion is absurd. Motion according to Descartes just is the '*transfer of one piece of matter, or one body, from the vicinity of the other bodies which are in immediate contact with it, and which are regarded as being at rest, to the vicinity of other bodies'* (AT 8A:53; CSM 1:233), and so is nothing other than a mode of extension. Modal distinctions allow separability in one direction: for any mode M of a substance S, S can exist without M, but not conversely. But then what is true for motion should equally be true for force. The imagined scenario, in which extension is conceived of as existing in the absence of motion and force, is consistent with both force and motion being modes of extension. As modes, force and motion would be modally distinct from the extended substances of which they are modes.

On the issue of how motion, since it is a mode, is transferred in collisions, Descartes's (AT 5:403–5; CSMK 381–2) August 1649 reply to More suggests that he thinks not that motion literally 'transmigrates' from one body to another, in the way that one might hand over a ten-dollar bill to someone else, but that there is an increase in the power of one body to move *in virtue of* a corresponding decrease in the body with which it has come into contact. A better analogy for this kind of 'transfer' of power might be how I increase your purchasing power by dropping my bid, even though no money changes hands. That Descartes thinks of transfers of motion in terms of coordinated and mutually dependent changes in the modes of bodies in contact is suggested by the following passage:

> You observe correctly that 'motion, being a mode of body, cannot pass from one body to another.' But that is not what I wrote; indeed I think that motion, considered as such a mode, continually changes. For there is one mode in the first point of a body A in that it is separated from the first point of a body B; and another mode in that it is separated from the second point; and another mode in that it is separated from the third point; and so on. But when I said that the same amount of motion always remains in matter, I meant this about the force which impels its parts, which is applied at different times to different parts of matter in accordance with the laws.
>
> (AT 5:404-5; CSMK 382)

This passage suggests that the transfer of motion as one body peels away from another in a collision consists in changes in their respective 'inclinations to move', and so, in turn, in their changing quantities of motion. Strictly speaking, it is motive force that is conserved in collisions, and only indirectly motion.[10]

---

[10] Whether this is a satisfactory explanation is another matter. What, you might ask, is the mechanism that ensures the coordinated increases and decreases of motion in colliding bodies? Is it God or must Descartes acknowledge something like Leibniz's pre-established harmony? Cf. Leibniz 1875–90, 4:520; 1997, 205.

Allowing that force, like motion, is a mode of bodies does not, of course, answer the question of what force is. Nor does it address other concerns that La Forge has concerning how it is that moving force, which seems to Descartes undeniably something that spiritual substances can also exert, is predicable of two such radically different kinds of substances. What the experiment does suggest, however, is that even if force is not the same thing as motion, the two notions are inextricably connected. This, however, tells us little about what exactly the relationship between them is. If a body at rest can exert or be subject to a force, then that force cannot depend necessarily on *its* being in motion. If force is merely *supervenient* upon matter in motion, it must be *global* rather than local to the body itself—that is, supervenient upon the distribution of motion in the universe. Indeed, this seems to be what Descartes has in mind when he describes light as pressure. The motion of a distant star, for example, produces an instantaneous force—the 'first preparation for motion'—that is applied through the intervening matter pressing against the retina of a perceiver, even though that matter touching the retina is not itself moving. 'It must be noticed that the force of light does not consist in any duration of motion, but only the pressing or first preparation for motion, even though actual motion may not result from this pressure' (AT 8A:114; Descartes 1982, 117). Light that consists in pressure is properly referred to as an 'action', for 'action is quite general: it comprises not only the power or inclination to move but also the movement itself' (AT 2:204; CSMK 109).

That there is matter in motion *somewhere* in the universe seems implied also by the very concept of a 'resisting force'. If all matter in the universe were at rest and *incapable* of motion, it would make little sense to suppose that anything exerts a resisting force. Resisting forces resist changes of state caused by bodies in collisions. In a motionless universe, the idea of a resisting force would be meaningless. But still, a body at rest exerts a resisting force.

Given the evidence adduced so far, it should seem obvious that Descartes is committed to the idea of active forces in nature. Why then has this seemed so obviously wrong to so many of Descartes's interpreters? Certain statements from his own hand, particularly those relating to the third law, seem to create the impression that he thinks that there is no activity in nature. As he writes to More in August 1649, '*matter left to itself and suffering no impulse from anything else is clearly at rest...however, impelled by God, it has as much motion or translation conserved by him as was laid down from the beginning*' (AT 5:404; CSMK 381; translation altered). As Alan Gabbey (1971/2, 12–13) has pointed out, it was far from clear that if matter was truly inert, it could comply with the conservation law. Both Beeckman and Newton realized that without active forces there would be no grounds for the conservation of motion. As early as 1620, Beeckman (1939–53, 2:45) argued that the motion of a body in a vacuum would decrease, not increase, without an active force to sustain it, and that in equal collisions where no one body surpasses the other, neither body would be deflected or transfer its motion to the other. In equal collisions, bodies should tend towards rest. Descartes's first collision rule from *Principles* II.46 (AT 8A:68) predicts that bodies in this scenario would spring apart. But what accounts for this elasticity if bodies are inert?

This is only one of many conceptual and interpretative puzzles. Descartes's referring to God's role in creating and conserving motion in explicating the third law adds to the impression that divine action has been substituted for natural causal powers. Yet, Descartes often refers to forces both 'of acting and resisting' (AT 8A:66; CSM 1:243) and attributes them to bodies. In a 28 October 1640 letter to Mersenne, in which Descartes criticizes an opinion of Father Lacombe, he writes:

> He is right in saying that it is a big mistake to accept the principle that no body moves of itself. For it is certain that a body, once it has begun to move, has in itself for that reason alone the force [*la force*] to continue to move, just as, once it is stationary in a certain place, it has for that reason alone the power to continue to remain there. But as for the principle of movement which he imagines to be different in each body, this is altogether imaginary.
>
> (AT 3:213; CSMK 155; translation altered)

In *Principles* II.25, he (AT 8A:53–4; CSM 1:233) then invokes a distinction between motion and 'the force or action' which brings about motion. And in an August 1641 letter to Hyperaspistes he explicitly resists the idea that bodies are only acted upon and do not act. In accounting for the continued motion of a top once the whip has been removed, Descartes attributes to the top the capacity to *act on itself* to sustain itself in motion. Rejecting this analysis would, Descartes argues, be tantamount to asserting that there could be potentialities without actualities and that everything that occurs in nature is simply the effect of the first act of creation, a consequence he takes to be absurd: 'Nor do I see why we could not as well say that there are now no activities in the world at all, but that all the things which happen are passivities of the activities that were there when the world began' (AT 3:428; CSMK 193).[11] But if matter is inert, aren't these exactly the conclusions a mechanist like Descartes ought to draw?

---

[11] This statement is not incompatible with occasionalism because all the 'activities in the world' could reflect God's *continual* activity. But this reading does not sit well with Descartes's attributing to bodies like the spinning top the capacity 'to act on itself'.

Here perhaps is where our understanding is easily derailed. The law of inertia commits us to the idea that bodies are *inertial*, not that bodies are *inert*. The two are distinct claims and Descartes's usual way of expressing the inertial quality of bodies is in the active voice. The idea that bodies exhibit tendencies to move is part of his explication of the idea that every body exhibits a *conatus*—a 'striving' to move in a specific direction given its circumstances whether the body is actually moving or not, and which, given the opportunity, accounts for the first moment of change from rest to motion (AT 8A:66; CSM 1:243). This striving is, as we saw in the account of light, 'the first preparation for motion', not motion itself. A body's striving persists even when it is at rest, and hence, Descartes concludes, striving is not the same thing as motion. In *Principles* II.43 Descartes equates the idea of *conatus* with the notion of force, in *Principles* II.25 he distinguishes it from motion, and elsewhere he identifies it as a mode and claims that it is conserved in collisions (AT 5:404; CSMK 381–2). Since the idea of a body's conatus or force includes the idea of its *directionality*, the discussion suggests, contrary to Descartes's at times overly narrow definition of the third law, that it is not just motion but also directionality that is conserved.[12]

Descartes speaks of the force of a body as measurable from its size, surface area, speed, and 'nature of its movement', and from the different ways in which other bodies encounter it (AT 8A:66–7; CSM 1:243–4). But this should not be taken as a complete reduction of the notion of force or power, for it is a certain kind of force that determines the *directionality* of the body and nothing in this list refers to directionality. Compare 'I add nothing about the direction in which each part moves. For the power to move and the power that determines in what direction the motion must take place are completely different things and can exist one without the other' (AT 11:8–9; CSM 1:83). But what *is* conatus or force and how is it consistent both with the rejection of occult qualities and with the general view that all changes in nature (not initiated by a mind) must be explicable in terms of extension and its modes?

2. Laws and Tendencies

Notice that Descartes's laws are *tendency laws*: they state what would happen if nothing interferes, not what always or even usually happens, and, since interference is inevitable, what they describe are tendencies. In accordance

---

[12]   See *Principles* II.40 (AT 8A:65; CSM 1:242) and his August 1649 comments to More (AT 5:403–5; CSMK 381–2). If Descartes did intend that directionality is conserved, it is not an idea he develops in his physics, leaving him open to Leibniz's objections. For discussion see Garber 1983, 106.

with the laws, bodies exhibit tendencies that may never be fulfilled and yet are causally relevant to explaining their behaviour. Let us consider each law in turn:

> First law: That each and every thing, in as much as it is in itself [*quantum in se est*], always perseveres in the same state; and thus what is once moved, always continues to move. (AT 8A:62; CSM 1:240–1; translation altered)

The operative phrase in the first law is 'in as much as it is in itself' and the rationale offered is that 'nothing carries [*ferri*] itself by its own nature towards its opposite, or towards its own destruction' (AT 8A: 63; CSM 1:241). Later, at *Principles* II.43, the ideas of force (the power…of acting or resisting; *vis…agendum vel resistendum*), the intrinsic powers of a body (*quantum in se est*), and tending (*tendat*) are brought together: the 'force of each body to act against another or to resist the action of another' consists 'in that each and every thing tends, in as much as it is in itself [*quantum in se est*], to remain in the same state' (AT 8A:66; CSM 1:243; translation altered).

The first law describes the tendency of a body to persist in the same state of motion or rest, which, given the inevitability of collisions in a plenum, is inevitably thwarted. The active language that Descartes uses in explicating the first law suggests, contrary to what many interpreters have thought, that he did not think that bodies are completely passive.

Consider now the second law:

> Second law: That every motion is in itself rectilinear; and hence any body moving in a circle always tends [*tendere semper*] to move away from the centre of the circle which it describes. (AT 8A:63; CSM 1:241; translation altered)

Bodies moving in a circular direction tend to move along the line of the curve (centripetal force) and in a straight line away from the centre (centrifugal force). Given that all bodies move in a circle (more precisely, a vortex), as a matter of fact, the centrifugal tendency is always naturally impeded. The centrifugal tendency is, however, causally relevant since it explains why bodies moving in a circle tend to move away from the centre.

Finally, the third law:

> Third law: If a body collides with another body that is stronger than itself, it loses none of its motion; but if it collides with a weaker body, it loses a

quantity of motion equal to that which it imparts to the other body. (AT 8A:65; CSM 1:242)

The third law states a fact—that motion is conserved—but relies on a distinction between the motion of a body, considered in itself, and the 'determination in a certain direction'. The cause of motion is not necessarily the same as the cause of the body's directionality. When a projectile is deflected upon collision with a larger body, it can change direction without its quantity of motion being affected. This change is attributed to the 'resistance of the body which deflects its path', which Descartes suggests in the collision rules is a function of the relative sizes and hardness of the colliding bodies and the density of the medium (AT 8A:71 CSM 1:245). The results of any collision depend on how such contests among forces play out.

In each of these laws, bodies are described as having tendencies—to persevere in the same state, to move in a straight line, and to change direction or not without affecting the overall quantity of motion in the universe—which are conceived of as causally relevant even when they are unfulfilled. Each tendency relates to bodies that either are in motion or could be—they are tendencies to move in certain directions or to resist being moved by other impacting bodies.

What reason do we have, though, for thinking of these tendencies as real? In their translation of Descartes's *Principles of Philosophy*, Miller and Miller (Descartes 1982, 113 n55) write in a footnote: '*a tendency that is never realised is a strange sort of tendency.*' For those who think that unrealized tendencies are incoherent, Descartes's tendency talk will just seem to be another *façon de parler*—shorthand perhaps for descriptions referring only to divine action. I hold to the contrary that there is no reason not to accept that Descartes thought of these tendencies or 'strivings' as real, as irreducible to other modes of bodies, and as predicable of bodies themselves. In particular, as we shall see, tendencies are *measurable* even when unfulfilled. None of this rules out that they are not, for all that, peculiar features of Descartes's physics.

3. Force and Potentiality

On 5 October 1637 Descartes writes to Mersenne that whether something is in actuality or potentiality, the same laws apply, for 'although it is not always true that what has once been in potentiality is later in actuality, it is impossible for something to be in actuality without having been in potentiality'

(AT 1:451; CSMK 74). Despite his antagonism towards Aristotelian philosophy, Descartes was thus not entirely averse to the idea of real potentialities or to the idea that potentialities are causally relevant. Indeed, given the way his laws are defined, tendencies can be seen to play many of the roles in Descartes's physics that potentialities played in Aristotle's. Both are causally relevant and explanatory even when unfulfilled. It is *because* of its centrifugal tendency that a body recedes from the centre of a circle; it never *actually* moves in a straight line, but that it tends to is a cause of its moving as it does.

Descartes takes the existence of tendencies to be experimentally 'confirmed' (AT 8A:111). In a critical set of passages from *Principles* III.58–64, the verbs '*tendere*' (to stretch out, to extend) and '*conari*'[13] (to endeavour, to strive) are used interchangeably. The stone in a sling discussed at *Principles* III.56 exhibits multiple, conflicting tendencies: centripetal and centrifugal tendencies and tendencies (resisted by the sling) to shoot off along tangents from points on the arc it describes. We are not concerned here with whether Descartes correctly describes the tendencies of the stone. What is important is that he means 'stretching' quite literally—the tending of a body in a certain direction creates a 'tension' (*tensio*), in this case in the sling, which is the measure of the body's force. At *Principles* III.59, he writes: 'this tension [*tensio*] having arisen by the sole force by which the stone strives to recede from the centre of its motion shows us the quantity of the force' (AT 8A:112). Forces, understood as tendencies, are, therefore, real and measurable for Descartes.

If we accept that tendencies are real, then we can make sense of other claims that Descartes makes about force. Tendencies are 'the first preparation for motion'—motive forces but not themselves motions (AT 8A:115). Because forces exert an influence even when bodies are at rest, when the obstacle to their moving is removed, they *instantaneously* move. We can see how a body's tendency to move in a straight line means that it will choose the shortest path if we consider the case of fermenting grapes in a vat which has two plugs in the bottom. The parts of wine will exhibit tendencies towards the closest hole: 'The parts of wine at one place tend to go down in a straight line through one hole *at the very instant* it is opened, and at the same time through the other hole, while the parts at other places also tend to go down through these two holes, without these actions being impeded by each other or by the resistance of the bunches of grapes in the vat' (AT 6:87; CSM 1:154; emphasis added). The

---

[13] Interestingly, some uses of *conari* mean 'to set oneself in motion', which alludes both to the role *conatus* plays in accounting for the first motion of change and the capacity of a thing to act upon itself.

insertion of the clause 'at the very instant' in the above passage is illuminating. It cannot be motion that explains how wine in the vat starts to move once the plugs are removed. Motion (the traversing of distance) takes time and cannot, therefore, be an efficient cause *at an instant*. Whereas motion is impeded by the resistance of other bodies, a tendency to move is not, which is why a tendency can be a cause where motion cannot.

Tendencies must be causally relevant for Descartes not only in accounting for the first instant of change, but also in explaining what happens at the instant of impact. The impact of a body is determined inter alia by its force of motion and force of directionality at the instant of impact. Whether one body at the point of impact with another springs back or carries the other body forward with it and transfers some of its motion to it is determined not just by the body's relative mass, surface area, and speed prior to collision, but also by its force at the moment of impact. Since the idea of motion at an instant is incoherent, it cannot be that motion substitutes for force in explaining what happens at the point of impact.[14]

Finally, and perhaps most importantly, Descartes is apt to use not just the terminology of motion and rest but also that of 'action' and 'passion', 'agent' and 'patient'. This terminology is needed in cases where forces are operative and observable even though the body is stationary. Consider the following tug-of-war or 'stalemate' case in which Descartes suggests to Morin (12 September 1638) that it is possible to distinguish the kind of force or action in operation (push or pull) even when the body to which the force is applied is motionless:

> For example, say two blind men are holding a stick and pushing it with equal force against each other, so that the stick does not move at all; each then pulls the stick with equal force towards himself, and again the stick does not move. In each case one man is exerting a force while the other is doing the same in the opposite direction; and the forces are so exactly equal that the stick stays motionless throughout. It is certain that each blind man, simply from the fact that the stick is motionless, can feel that the other is pushing it or pulling it with equal force. What each man feels in the stick, namely a lack

---

[14] Valicella (1999, 11–12) cites Tooley's and Armstrong's objections to Russell's attempt to ground vector properties at an instant in categorical properties such as the occupation of different places at different times. Armstrong (1997, 77) introduces the case, attributed to Bigelow and Pargetter, of a meteor crashing into Mars and causing a crater: 'The size of the crater is proportional to the force exerted on impact, together, of course, with whatever Mars happens to be doing at that instant. But if the force is not something, a property presumably, that the meteor has at the moment of impact, then how is the particular size of the crater to be explained?'

of movement in various respects, can be called various actions which are impressed on it by the various exertions of the other man. For when one of them is pulling the stick, this does not cause the other to feel the same action as when he is pushing it.   (AT 2:363; CSMK 120–1)

Although Descartes distinguishes between the 'moving body' and the 'moved body', this distinction is insufficient to capture the distinction between 'acting' and 'being acted upon'. Given the *relativity* of motion, there is no categorical difference between a moving and a moved body, whereas there is a categorical difference between an agent and a patient and, as the above passage suggests, categorical differences between the types of actions being performed. In any transaction there is a fact of the matter which body is exerting an acting force and what kind of force it is. It is not a matter of perspective, for example, whether my car has a tendency to pull to the left or the rest of the universe a tendency to pull to the right around my car. But if motion and rest are relative, it is just as true to say that the universe is moving to the right of my car as it is to say that my car is moving to the left. In the tug-of-war example, there is a categorical difference between the push-force and the pull-force, a fact that is illustrated by the different counterfactuals each supports. If A is pushing the stick and B suddenly stops pushing back, the stick will move towards B; if A is pulling and B stops pulling, the stick will move towards A. Since the stick's properties are the same in each scenario—same mass, surface area, size, and state of rest—it can only be the different kind of action or force that explains which counterfactual is true in the situation.

This is important for it shows that Descartes is more of a realist about tendencies than some contemporary philosophers would think is necessary. In contemporary debates about tendencies, some are inclined to analyse tendency talk in terms of the truth of subjunctive conditionals.[15] To say that the wine has a tendency to move towards the hole in the bottom of the vat just is a way of expressing the truth of the conditional statement that if the plug were removed, the wine would flow out of the hole in the bottom of the vat.

[15] For an overview of contemporary views on tendencies, see Gibson 1983, 300–2. On the subjunctive view, an unfulfilled tendency is analysed by reference to closest worlds or situations where the tendency is fulfilled. There is a certain amount of vagueness in such views in specifying which worlds are closest and appealing to possible worlds to avoid referring to real tendencies seems to carry little ontological gain. Gibson's own preference is for a view that holds tendency statements to be partial truths, analysed in terms of conditionals in which the antecedent is left unspecified. 'The car has a tendency to move to the left' on this view just means that under certain conditions (only partially specified, for example, by reference to some sample conditions) it invariably moves to the left. The presence of a suppressed antecedent does not alleviate the vagueness attending the subjective view but seems only to add to it.

But for this to suffice, an account of the categorical basis of tendencies in Descartes's physics would require that Descartes is committed to counterfactuals being *primitively* true, and there is no evidence that he would find such a position tenable. It would violate the truthmaker principle to which he generally adheres.[16] The truth of the relevant subjective conditionals is explained by reference to tendencies, not conversely:

> And note here it is necessary to distinguish between the movement and the action or tendency to move. For we may very easily conceive that the parts of wine at one place should tend towards one hole and at the same time towards the other, even though they cannot actually move towards both holes at the same time, and that they should tend exactly in a straight line towards one and towards the other, even though they cannot move exactly in a straight line because of the bunches of grapes which are between them.
>
> (AT 6:88; CSM 1:155)

If tendencies are truthmakers for counterfactuals, then we should expect them to be assigned the same causally explanatory roles that dispositions are assigned. We see some evidence for this in other texts. The following passage from *Principles* II.23 attributes 'all the variety in matter' to the *capacities* of bodies to undergo certain kinds of changes—their 'divisibility' (*partibilis*), 'mobility' (*mobilis*), and 'capacity to be affected' (*capax...affectionum*): 'All the variety in matter, or all the diversity of its forms, depends on motion.... All the properties which we clearly perceive in it are reducible to its divisibility and consequent mobility in respect of its parts, and its resulting capacity to be affected in all the ways which we perceive as being derivable from the movement of the parts' (AT 8A:52; CSM 1:232). The dispositional language in this passage suggests that Descartes is not averse to the idea that there are real dispositions. Elsewhere, he appears happy to refer to dispositions, provided that they are not thought to be really distinct from substances (bodies or minds) and thus substances in their own right. Dispositions are as much in the category of modes as actual properties of things:

> Now we do not deny active qualities, but we say only that they should not be regarded as having any degree of reality greater than that of modes; for to

---

[16] As Descartes writes in the Fifth Meditation, 'All these properties [of a triangle] are certainly true, since I am clearly aware of them, and therefore they are something, and not merely nothing; for it is obvious that whatever is true is true of something' (AT 7:65; CSM 2:45).

regard them so is to conceive of them as substances. Nor do we deny dispositions [*habitus*] but we divide them into two kinds. Some are purely material and depend only on the configuration or other arrangement of the parts. Others are immaterial or spiritual, like states of faith and grace [*habitus fidei, gratiae*] etc....; these do not depend on anything bodily, but are spiritual modes inhering in the mind, just as movement and shape are corporeal modes inhering in the body.   (AT 3:503; CSMK 208)

Bodily dispositions are secondary to particular movements and configurations of parts of matter.

The categorical distinction between agent and patient may help to alleviate the problem that what happens in collisions is not necessarily frame-invariant. A large body, A, hits a smaller one, B, and transfers some of its motion to it.[17] Holding the relative sizes constant, however, if the smaller body B hits the larger one A, Descartes predicts that something different happens—the smaller body B retains the same quantity of motion and is deflected. In both scenarios, the same quantity of motion is conserved but with different distributions. Yet the two scenarios look symmetrical—the same collision described first from the point of view of the larger body and then redescribed from the point of view of the smaller body. So what exactly accounts for the different distributions of motion? If motion and rest are relative, we should be able to predict exactly the same outcome no matter which way the scenario is described. One possibility is that there are different forces at play, analogous to the different push and pull forces described in the tug-of-war case above. If A is pushing and B fails to resist, A will continue moving towards B and push it along, whereas if B is pushing and fails to overcome A, B will bounce back. Given the difference in forces, different counterfactuals come into play and the two scenarios are not, despite appearances, symmetrical.

4. Real Tendencies and Final Causes

The above discussion points to the conclusion that Descartes equated the notion of 'force' with that of 'tendency' and thought of tendencies as

---

[17] This tension is evident in Descartes's 4th and 5th collision rules (*Principles* II.49–50). Schmaltz (2008, 89) argues that Descartes needs a non-relativistic notion of motion and rest to account for these differences. See also Garber (1992, 240–1) on the incompatibility between these two rules and the arbitrary distinction between motion and rest.

real, causally relevant, and irreducible to other modes of extension. One question that remains to be answered is whether tendency talk is inherently teleological. Is describing a body as having a tendency to move or a tendency to move in a certain direction tantamount to attributing to it an end or purpose?

A positive answer would be inconsistent with Descartes's denial of final causes. Descartes objects to ends or final causes having any role in physics, first, because it is hubris to assume one can know the purposes of God (AT 7:55; CSM 2:38; AT 8A:81; CSM 1:248; AT 4:292; CSMK: 266); second, because God's intentions are just another kind of efficient cause (AT 8A:15–16; CSM 1:202); and, finally, because if we suppose that bodies act for the sake of ends, then we must suppose them capable of cognition, which is absurd for non-thinking, extended substances. This last objection constitutes the basis of Descartes's objection to the scholastic notion of *gravitas* (heaviness) on the grounds that it confuses the way in which minds move bodies with the way in which bodies move (AT 7:442; CSM 2:298): 'But what makes it especially clear that my idea of gravity was taken largely from the idea I had of the mind is the fact that I thought gravity carried bodies towards the centre of the earth as if it had some knowledge of the centre within itself. For surely this could not happen without knowledge, and there can be no knowledge except in a mind' (AT 7:442; CSM 2:298; see also AT 3:667–8; CSMK: 219). Instead, the laws of nature suffice to explain all the interactions among bodies without the need to make assumptions about ends or divine purposes: 'I can conceive such [natural] inclinations only in things which possess understanding, and I do not attribute them even to animals which lack reason. Everything in them which we call natural appetites or inclinations is explained on my theory solely in terms of the rules of mechanics' (Descartes to Mersenne, 28 October 1640, AT 3:213; CSMK: 155). Descartes extends this thought even to describing the inclinations of organic bodies. An animal body has no more thought about the direction to which it tends when it is hunting prey, seeking a mate, or fleeing a predator than a clock has thoughts about time (AT 11:202; CSM 1:108; AT 11:331; CSM 1:329–30; AT 6:55; CSM 1:139; AT 5:278; CSMK: 365; AT 3:566: CSMK: 214).[18] That tendency talk does not import teleological assumptions back into natural philosophy is an assumption Descartes shares with Hobbes, to whom we now turn.

---

[18] See Brown 2012 for an account of how to render Descartes's use of functional attributions consistent with his denial of final causality.

5. Hobbes on Tendency and Motion

The lack of any clear categorical basis for Descartes's talk of 'forces of acting and resisting' remained a thorn in the side of Cartesians, impelling many in the direction of occasionalism until Leibniz and Newton made primitive forces fashionable again. Attracted to neither of these solutions, Hobbes pursued a different course, reducing 'force' to a species of motion that could, he thought, be effective at an instant.

Hobbes equates the ideas of cause and power, observing that the only difference is that when we speak of causes we tend to speak of what is past, whereas powers always pertain to the future (Hobbes 1962, 1:127–8). Hobbes's laws too are 'tendency laws'—were the *conatus* (endeavour) of a body made in empty space, he argues, it would proceed with the same velocity indefinitely (1:216–17). The obvious difficulty for Hobbes's conflation of 'force' and 'motion' is accounting for the force of a body at rest. But this is a bullet Hobbes is prepared to bite. When a body opposed by some stronger force is subsequently at 'rest', it is nonetheless in a state of 'infinitesimal' and 'imperceptible' motion. The reason why Hobbes conflates tendencies and motions is simply that he cannot envisage any kind of change which does not consist in motion because all change involves continual progression—'a continual mutation in the agent and patient'—which, being divisible into parts, requires us to think of each part, including the beginning, as an action and cause of motion. We cannot conceive of a natural change except as something affected in our senses, but there are no changes in sensation that are not the product of motion, either as the result of a change in motion in the agent or a change of motion in the patient. Completely inactive rest would thus be neither a cause of anything nor something we could really comprehend (1:126; 1:131). But Hobbes is also concerned to explain how a body *begins* to move out from a state of rest. If, in order to *begin* to move requires traversing a space, even an infinitesimal one, how could a body not in motion already ever begin to move? 'For let a space be never so little, that which is moved over a greater space, whereof that little one is part, must first be moved over that' (Hobbes 1968, 119). Since only a body already in motion can produce motion or, for that matter, resist motion when another body is pressing against it, the conatus of a body, Hobbes concludes, must be a species of motion (Hobbes 1962, 1:123–4).

Velocity is thus simply 'motion considered as power'—the power by which a body moved may in a certain time transmit a certain length (1:203). The body's conatus or striving is its *instantaneous* velocity: '*motion made in less space and time than can be given; that is, less than can be determined or assigned by exposition or number; that is motion made through the length of a point and an instant of time'* (1:206). Many of the roles played by actions, forces, or tendencies in Descartes's physics—accounting for the first moment of change, accounting for the force at impact, and serving as the truthmaker for counterfactuals that express the effects of the laws of mechanics—are performed by Hobbes's notion of infinitesimal motion.

Hobbes's infinitesimals are real, even if only relatively defined. An infinitesimal point, Hobbes writes, is a determinate body, of which length, breadth, and depth are ignored (7:202). A point is *'a quantity less than any quantity that can possibly be named'*, just as a line is a body of which breadth and depth are ignored and a plane, a body of which depth is ignored (1:211). Although infinitesimal, not all endeavours or forces are equal. We can conceive of differences between them much as we can conceive of differences in a point depending on the different angles terminating in them, and we can measure the effects of different endeavours or tendencies by comparing their velocities (1:206–7). A body's impetus is the measure or quantity of its endeavour (1:206–7), and its force is calculated from the impetus and magnitude of the moving body, *'whereof the said movent works more or less upon the body resisting it'* (1:212).

In summary, Hobbesian tendencies are real, causal powers that are neither dispositions nor potentialities. They may seem, therefore, to have significant advantages over Cartesian tendencies. Since they belong in the category of motion, there is nothing particularly occult or mysterious about what they are or how they exert a causal influence. Nor does appealing to infinitesimal motions introduce worries about final causation or teleology. Endeavours are as blind to their effects as any other motion of a body within the mechanical framework. The main problem with Hobbes's account of force or tendency is that it fails to preserve a categorical distinction between the agent and the patient.

The difference between motion and rest is, for Hobbes, reduced to a difference in degree rather than kind. A body at rest is still a body in motion, just infinitesimal motion. If motion and rest are relative notions, and if what we call a body's 'action' is indistinguishable from its motion, then whether a body is the agent or the patient in an interaction is also relative to our frame of reference. The categorical difference between agents and patients, which on Descartes's view proves indispensable to understanding what happens in collisions, is, in Hobbes's framework, collapsed into the relative notions of motion and rest. This, as we shall see, creates logical problems for Hobbes's account of what happens in cases where there is a contest of equally powerful forces.

6. Stalemate Cases

Why is it so important to preserve the distinction between agent and patient? Reconsider the stalemate case, a case in which a body is held stationary between two equally opposing forces or endeavours—for example, the stick suspended between the two blind men exerting equal push or pull forces or an iron ball suspended between two equally powerful magnets. Given the reduction of force to motion and Hobbes's definition of motion—viz. *'the continual relinquishing of one place and acquiring of another'* (Hobbes 1962, 1:109; 1:206–7; 1:211–13)—it follows that the 'stationary' body in a stalemate situation must be moving in opposite directions at once. But how is this possible?

Hobbes's identification of force and motion appears to lead him straight into a dilemma. He must either accept that a body subjected to equally powerful opposing forces is in motion and so indeed traverses a distance in opposite directions at once, or he must deny that any distance is traversed at all. The first option is absurd, and to assert the second, Hobbes must relinquish either the claim that endeavour is motion or the definition of motion as the traversing of distance over time. Neither of these is a palatable option. If a body is not in motion, not even infinitesimal motion, then, by Hobbes's own definition, it cannot be the subject of any endeavour. And the idea of motion being anything other than the transfer of a body across a distance over time is incoherent.

There is perhaps a way that Hobbes might have responded to such an objection and that is by denying that traversing an infinitesimal distance an infinite number of times ever amounts to traversing a finite distance. If an infinite number of infinitesimals is still infinitesimal, then the fact that a single body is infinitesimally moving in opposite directions at the same time would not perhaps produce the paradoxical result of a body traversing a finite distance in opposite directions at once. But aside from whatever other mathematical problems such a suggestion would have caused for Hobbes, the problem lies in accounting for how it is that through having an endeavour (that is being in a state of infinitesimal motion) a body ever moves any finite distance. Suppose a body has to move a certain distance—one metre, say, per second. If the following four conditions are met—(i) motion is the ratio of distance to time, (ii) one second is composed of infinitely many moments,

(iii) at each moment a body covers an infinitesimal distance, and (iv) an infinite infinitesimals is still infinitesimal—then no distance will be traversed no matter how much time is allowed.[19]

Hobbes's geometrical thinking did not incline him to think of points and lines as dimensionless—they have extension, just smaller than can be measured. And his insistence that any body that moves over a greater space must first be moved over a smaller one, in conjunction with his claim that endeavour is the first preparation for motion, suggests that he would have to allow that a body engaged in infinite infinitesimal motions covers a finite distance. But then, stalemate cases pose an intractable problem for his view.

7. The Mind's Tendencies to Move Matter

One reason for adopting occasionalism was, as we noted earlier, that the alternative is to suppose that there is an unequivocal notion of 'force' or 'tendency' that is predicable of both extended and spiritual substances. What, you might ask, could such a mode be? An attribute?[20] There are, after all, a number of generic attributes, such as number and duration, which are attributes of both extended and spiritual substances. Why shouldn't the power or force to move bodies similarly be accepted as an attribute or mode that belongs to either extended or spiritual substances?

One counter-intuitive consequence of this laissez-faire position is that, if the power to move bodies is an intrinsic attribute of mind, then we should be able to derive the concepts of body and extension from the idea of mind, which Descartes denies. Such a conclusion would be dire for it is contrary to a basic precept of Descartes's dualism that the concept of mind *excludes* (and is not merely abstracted from) the concept of body (AT 7:227–8; CSM 2:160).

---

[19] There are, one might object, always going to be problems associated with Zeno's paradox in explaining how a body travels along any line that is composed of infinitely many points. Even allowing that an infinite number of infinitesimal points traversed amounts to a finite distance underdetermines what the exact distance to time ratio is in any given instance. But clearly nothing is gained by *not* making this allowance.

[20] For different reasons, Manchak (2009) argues that force needs to be understood as an attribute, one that is not really but only conceptually distinct from extension. He is the only author whom I know countenances this possibility, but I find myself having to disagree. Manchak is led to construe force as a 'way of thinking' about extended substance (305), which I fear is based on a common confusion between Descartes's *distinctio rationis* and (merely) rational or 'conceptual' distinctions. The *distinctio rationis*, as Descartes uses it and as it would have been understood by his contemporaries, requires a foundation *in re* outside the mind and is applied by Descartes only to inseparable, unchanging attributes of substance, not to variable modes like forces. On the distinction of reason and Descartes's distinctions generally, see Brown 2011.

For a materialist like Hobbes, by contrast, the fact that 'power', 'force', and 'tendency' can be predicated univocally of minds and bodies poses no particular difficulty. The powers of the mind to move bodies are nothing other than special endeavours of the human body to persevere in existence in accordance with the first law. Descartes's *rationale* for the first law—that it is inconceivable that something tend toward its own destruction—is transformed in Hobbes's framework into the first principle of both nature and civil society: 'A LAW OF NATURE is a precept or general rule, found out by reason, by which a man is forbidden to do that which is destructive of his life or taketh away the means of preserving the same; and to omit that by which he thinketh it may best be preserved' (Hobbes 1968, 64.) Human endeavours just are the body's vital motion, which can be increased or decreased through environmental impacts. Increases in vital motions are experienced in conscious beings as appetites or desires for certain objects, whereas decreases are experienced as aversions, and it is these passions of desire and aversion that are the proximal motive forces behind all conscious human activity.

> This motion, in which consisteth pleasure or pain, is also a solicitation or provocation either to draw near the thing that pleaseth, or to retire from the thing that displeaseth; and this solicitation is the endeavour or internal beginning of animal motion, which when the object delighteth, is called appetite; when it displeaseth, it is called aversion, in respect of the displeasure present; but in respect of the displeasure expected, fear. So that pleasure, love, and appetite, which is also called desire are divers names for divers considerations of the same thing.   (Hobbes 1962, 4:31–2)

Deliberation and will are not forces autonomous from the forces of natural bodies, for Hobbes, but consist simply in the *alternation* of contesting forces or endeavours in the animal or human body. When we alternate between appetite and aversion, hope, fear, desire, despair, when we consider the pros and cons of a certain course of action, its possibly good or evil consequences, the probability or improbability of attaining what we desire, 'the whole summe of Desires, Aversions, Hopes and Fears, continued till the thing be done, or thought impossible, is that we call Deliberation' (Hobbes 1968, 127).

Human bodies too can find themselves in stalemate positions. Where we are rendered motionless by conflicting passions—for example, by equal measures of longing and despair—analogous problems will arise for Hobbes's view of human action as arise for his analysis of force as motion in physics. How such 'alternation' or 'deliberation' among the passions is to be understood at the mechanical level is unclear. When I am torn, like Buridan's ass, between two equally compelling goods, am I subject to an infinitesimal motion drawing me towards and away from each object at the same time, or to multiple endeavours located in different parts of my body that cancel out each other's effects? The first option is problematic for the reasons adduced above, namely, that it requires that a body traverse a distance however minute in opposite directions at once. While the second option makes more sense, particularly in the context of such a complex, functionally integrated composite as the human body, it puts pressure on Hobbes's account of how organic bodies are individuated. The very unity and identity of a composite body that is continually undergoing replacement of its parts is dependent upon a single, continuous endeavour that persists from its first moment of existence until its death. It is 'the unbroken nature of the flux by which matter decays and is replaced' that defines the identity of a living thing over time (Hobbes 1962, 1:206; 1976, 142). If this single, individuating endeavour is equated with a single motion of the body, then the problem of explaining how a body can be pushed or pulled in opposite directions at once remains.

The intrinsic power of the mind to move bodies on Descartes's view, by contrast, is made possible by the 'substantial union' of mind and body (AT 7:228; CSM 2:160) As he writes to Princess Elisabeth on 21 May 1643, 'As regards the soul and the body together, we have only the notion of their union, on which depends our notion of the soul's power to move the body and the body's power to act on the soul and cause its sensations and passions' (AT 3:665; CSMK 218). Taking Descartes at his word here, 'force' or 'the power to move a body' applies univocally to minds and bodies, but to minds only because they exist in union with a body. It is this fact that blocks the conclusion that, if forces to move matter are predicable of the mind, the concept of extension must be contained within the concept of mind. It is only in the concept of the union that the power of the mind to move matter is contained. Since the mind can exist apart from matter and so apart from the union, the fact that the concept of the union entails the capacity of the mind to move matter is no threat to Descartes's dualism.

This solution cannot, however, be extended to explain the sense in which 'force' is predicable of God since there is no sense in which God and bodies form a union. Nor will Descartes's usual treatment of force as a mode of body make sense in the context of a being who has no modes and is not subject to change. Despite Descartes's suggestion to Henry More that it is the idea of mind–body interaction that derives from my idea of the union that grounds my idea of how God moves matter, since 'no mode of action belongs univocally to both God and his creatures', the one idea, as Roger Ariew (1983, 36) has argued, can at best serve as an analogy for the other.

It is no disgrace to a philosopher to believe that God can move a body, even though he does not regard God as corporeal; so it is no more of a disgrace for him to think much the same of other incorporeal substances. Of course I do not think that any mode of action belongs univocally to both God and his creatures, but I must confess that the only idea I can find in my mind to represent the way in which God or an angel can move matter is the one which shows me the way in which I am conscious I can move my own body by my own thought.   (AT 5:347; CSMK 375)

What is referred to, therefore, as the force by which God moves matter can be nothing other than God's being the 'universal and primary cause, which is the general cause of all motions in the world' (AT 8A:61; CSM 1:240), the manifestations of which consist in changes in the tendencies of created things, not changes within the creator.

As in Hobbes's theory of affect, sensations and passions are, for Descartes, natural inclinations or tendencies that contribute to the preservation of the human organism. They 'depend absolutely' upon the movements of the animal spirits that produce them (AT 11:359–60; CSM 1:343), and their 'principal effect' is to 'move and dispose the soul to want the things for which they prepare the body' (AT 11:359; CSM 1:343). Apparent conflicts within the soul are analysed in terms of a contest between forces originating from within the body and volitions originating from within the soul. These forces compete for control over the pineal gland leading to its being pushed to one side and then to the other. '[The pineal gland] can also be moved in various different ways by the soul, whose nature is such that it receives as many different impressions—that is, it has as many different perceptions as there occur different movements in this gland' (AT 1:355; CSM 1:341). When these forces are equal, the subject will experience being pushed toward and away from a particular course of action at one and the same time and will not move, much as the blind men feel the competing actions operating on the stick even though it and they are not moving (AT 11:364–6; CSM 1:345–6). But because force is not conflated with motion in Descartes's philosophy, there is no problem analogous to Hobbes's in making sense of what happens when the contest of forces produces a stalemate. A motive force operating within a body can be counteracted by a force originating from within the soul, or vice versa, while the pineal gland inclines in neither direction.

Contrary to La Forge's suspicions, therefore, there is no obvious dilemma produced by supposing that 'power' and 'force' can in Descartes's system be predicated of distinct kinds of substances, mind and body, despite the fact that the categorical basis in each case is radically different. This is what we should expect of a 'true union' of mind and body, which is not to say that there are not other problems associated with the idea of immaterial substances contributing forces to what otherwise appears to be a closed physical system.

8. Conclusion

As the notion of 'force' entered the theoretical physics being developed in the early part of the seventeenth century, mechanical philosophers clearly struggled to understand its role and its relationship to extension. Descartes and Hobbes stand as figures united in a common endeavour to demonstrate the power and universality of the mechanical explanation of nature but forced to disagree about one of its more fundamental concepts. In light of the difficulties faced in rendering a notion of force consistent with mechanical principles, occasionalism could well have appeared a simpler, more coherent, and theologically sanctioned choice. But relying upon God's direct, efficient causality to explain every change in nature would have rendered the foundations of the new physics ultimately unknowable. Descartes and Hobbes were too clever or too arrogantly confident in their own abilities as scientists to take the path least resisted. Understood as measurable tendencies, forces can have no other reasonable subject than bodies themselves.[21]

References

Ariew, Roger. 1983. 'Mind-body interaction in Cartesian philosophy: A reply to Garber'. In 'The Rationalist Conception of Consciousness'. Supplement, *Southern Journal of Philosophy* 21, no. S1 (Spring): 33–7.

[21] I am deeply indebted to the Rotman Institute of Philosophy, Western University, and to the organizers of the Rotman Summer Institute on Causal Powers, 2014, and editors of this volume, Benjamin Hill, Henrik Lagerlund, and Stathis Psillos, *causae sine qua non*. I am also grateful to audiences of the Rotman Institute, the UCLA Early Modern Conference, 2015, and the 'Activity, Spontaneity and Agency in Later Medieval and Early Modern Philosophy' conference at the University of Toronto, 2016, for helpful discussion. My thanks, as always, to Calvin Normore for many illuminating conversations about power and passion.

Armstrong, D. M. 1997. *A World of States of Affairs*. Cambridge: Cambridge University Press.

Beeckman, Isaac. 1939–53. *Journal tenu par Isaac Beeckman de 1604 à 1634*. 4 vols. La Haye: M. Nijhoff.

Brown, Deborah. 2011. 'The Duck's Leg: Descartes's Intermediate Distinction'. *Midwest Studies in Philosophy* 35: 26–45.

Brown, Deborah. 2012. 'Cartesian Functional Analysis'. *Australasian Journal of Philosophy* 90 (1): 75–92.

Cartwright, Nancy. 1989. *Nature's Capacities and their Measurement*. Oxford: Clarendon Press.

Champlin, T. S. 1990–1. 'Tendencies'. *Proceedings of the Aristotelian Society* 91: 119–33.

Cudworth, Ralph. 1837–8 (1678). *The True Intellectual System of the Universe*. 2 vols. Ed. Thomas Birch. Andover: Gould & Newman.

Descartes, René. 1982. *Principles of Philosophy*. Ed. and trans. Valentine Rodger Miller and Reese P. Miller: Dordrecht: Kluwer.

Des Chene, Dennis. 1996. *Physiologia: Natural Philosophy in late Aristotelian and Cartesian Thought*. Ithaca: Cornell University Press.

Gabbey, Alan. 1971/2. 'Force and Inertia in Seventeenth Century Dynamics'. *Studies in History and Philosophy of Science* 2 (1): 1–67.

Garber, Daniel. 1983. 'Mind, Body and the Laws of Nature in Descartes and Leibniz'. *Midwest Studies in Philosophy*. 8: 105–33.

Garber, Daniel. 1992. *Descartes' Metaphysical Physics*. Chicago: University of Chicago Press.

Gibson, Quentin. 1983. 'Tendencies'. *Philosophy of Science* 50 (2): 296–308.

Grosholz, Emily. 1988. 'Geometry, Time and Force in the Diagrams of Descartes, Galileo, Torricelli and Newton'. *Proceedings of the Biennial Meeting of the Philosophy of Science Association* 2: 237–48.

Gueroult, Martial. 1980. 'The Metaphysics and Physics of Force in Descartes'. In *Descartes: Philosophy, Mathematics and Physics*, 196–229. Ed. Stephen Gaukroger. Sussex: Harvester.

Harvey, William. 1910 (1628). *Excertatio anatomica de motu cordis et sanguinis in animali*. Trans. Robert Willis. New York: Collier & Sons.

Hatfield, Gary. 1979. 'Force (God) in Descartes' Physics'. *Studies in History and Philosophy of Science* 10: 113–40.

Hobbes, Thomas. 1962. *The English Works of Thomas Hobbes*. Ed. William Molesworth. Darmstadt: Scientia Verlag Aalen.

Hobbes, Thomas. 1962. 1968. *Leviathan*. Ed. C. B. Macpherson. London: Penguin.

Hobbes, Thomas. 1962. 1976. *Thomas White's De mundo examined*. Trans. Harold Whitmore Jones. London: Bradford University Press in association with Crosby Lockwood Staples.

Leibniz, Gottfried. 1898 (1714). *The Monadology and Other Philosophical Writings*. Trans. Robert Latta. Oxford: Oxford University Press.

Leibniz, Gottfried. 1875–90. *Die Philosophischen Schriften von Gottfried Wilhelm Leibniz*. 7 vols. Ed. C. I. Gerhardt. Berlin. Weidman.

Leibniz, Gottfried. 1997. *Leibniz's 'New System' and Associated Contempory Texts*. Ed. and trans. Roger Woolhouse and Richard Francks. Oxford: Oxford University Press.

Malebranche, Nicolas. 1997. *The Search After Truth*. Trans. Thomas M. Lennon and Paul J. Olscamp. Cambridge: Cambridge University Press.

Manchak, John Byron. 2009. 'On Force in Cartesian Physics'. *Philosophy of Science* 76 (3): 295–306.

More, Henry.1925. *Philosophical Writings*. New York: Oxford University Press.

Ott, Walter. 2009. *Causation and Laws of Nature in Early Modern Philosophy*. Oxford: Oxford University Press.

Schmaltz, Tad. 2008. *Descartes on Causation*. Oxford: Oxford University Press.

Valicella, William F. 1999. 'God, Causation and Occasionalism'. *Religious Studies*, 35 (1): 3–18.

Westfall, Richard. 1971. *Force in Newton's Physics*. New York: Elsevier.

5

The Ontological Status of Causal Powers

Substances, Modes, and Humeanism

Benjamin Hill

Occasionalism and the considerations supporting it really should be part of the contemporary discussions regarding causal powers. Not the thesis of occasionalism, of course. *God is the only true efficient cause* is neither plausible nor relevant to today's metaphysical conversations—none of us should take the thesis of occasionalism all that seriously as an account of the metaphysics of causation. But the underlying worries and concerns that drove some philosophers in the seventeenth century to occasionalism is another matter altogether. *No physical object can truly be an efficient cause*, the negative thesis of occasionalism, was driven by a cluster of concerns that have not yet been clearly resolved. These concerns were by and large focused on explaining the causality of secondary causes, and it is here that they intersect with the current discussions regarding causal powers. In the seventeenth century, these concerns involved central metaphysical features of the Aristotelian conceptions of powers and substances. If we are to now return to a broadly Aristotelian metaphysics of causal powers, inherent potentialities, and primary substances, we should pay careful attention to the metaphysical concerns that led earlier thinkers to abandon this architectonic. If we now have proper resolutions to those worries, great! Let's hear them, and once and for all dispel these worries. But even if we do not have proper resolutions yet, we should have a clearer idea of the issues facing us and a firmer basis for assessing the prospects for proper resolutions.

The limited goal of this chapter is to initiate that critical reflection and conversation, to bring forth some of the worries and concerns that motivated the occasionalists of the seventeenth century, and flag the ways that they still seem relevant to today's metaphysicians regarding causal powers. Whether and how contemporary practitioners benefit from this, I will leave for others as they carry on the discussion about causal powers.

Benjamin Hill, *The Ontological Status of Causal Powers: Substances, Modes, and Humeanism* In: *Reconsidering Causal Powers: Historical and Conceptual Perspectives*. Edited by: Benjamin Hill, Henrik Lagerlund, and Stathis Psillos, Oxford University Press (2021). © Benjamin Hill.
DOI: 10.1093/oso/9780198869528.003.0006

Most of us begin our thinking about critiques of causal powers with Hume. That is not without reason—Hume's critique is an especially powerful, cogent, and elegant piece of philosophical writing. Plus, there is no greater wordsmith in the history of philosophy than Hume; he is always an absolute joy to read. And the crispness of his writing makes his thought sparkle, there is no question about that. One might think that Hume is sufficient and that there is no real need to consider the occasionalists' worries and concerns about Aristotelian causal powers. Hume's critique was, after all, strongly influenced by Malebranche's (McCracken 1983), and, so the thought goes, all the important features of the occasionalists' critiques are essentially contained in Hume's or can be easily reconstructed from it. I strongly disagree. Hume's critique is very different from the occasionalists', even from Malebranche's, despite the real and substantive influence Malebranche imposed on Hume's thinking, and there is a real danger that Hume's use of the occasionalists' ideas misrepresent their concerns. We don't want to be distracted by Hume's sceptical antics or rest contented with the belief that either Humean scepticism or a narrowly constricted, Humean empiricism is the primary basis for rejecting causal powers. So, it is important that we look at the occasionalists' critiques themselves and not view them only through the words of Hume.

Thus, I'd like to return us to the occasionalists' ideas by first considering Hume and Locke. Walking us back through Hume to Locke lets us see some parallels between Locke's position and the current neo-Aristotelian empiricism, which counters the idea that looking to Hume is sufficient for contemporary practitioners. It is all too easy to dismiss the Humean critiques as rooted in the scepticism engendered by a narrow and simplistic empiricism, and to believe that more mature, contemporary attitudes toward science and knowledge provide a firmer basis for grounding the postulation of causal powers. But Locke too failed to take the underlying metaphysical motivations for occasionalism seriously and focused his critical comments of it solely on the thesis itself. I shall bring out the underlying concerns that were driving occasionalist thinkers, especially with regard to their negative thesis, and emphasize the ways that they remain current. My fundamental fear is that contemporary practitioners also fail to take these underlying worries seriously and properly acknowledge or address them, with the result that they continue to fester and eventually lead to the return and re-entrenchment of radical Humeanism. It is out of concern for this that I offer these reflections on the nature of occasionalist thinking.

1. From Hume back to Locke

It is with good reason that Hume is the customary place for metaphysicians to begin reflecting on causal powers. Consider his powerful, sustained argument against causal powers inherent in bodies:

> It must certainly be allowed, that nature has kept us at a great distance from all her secrets and has afforded us only the knowledge of a few superficial qualities of objects; while she conceals from us those powers and principles, on which the influence of those objects entirely depends.... Suppose a person, though endowed with the strongest faculties of reason and reflection, to be brought on a sudden into this world; he would, indeed, immediately observe a continual succession of objects, and one event following another; but would not be able to discover any thing further. He would not, at first, by any reasoning be able to reach the idea of cause and effect; since the particular powers, by which all natural operations are performed, never appear to the senses;.... There are [therefore] no ideas, which occur in metaphysics, more obscure and uncertain, then those of *power, force, energy,* or *necessary connexion* ....To be fully acquainted, therefore, with the idea of power or necessary connexion, let us examine its impression;...When we look about us towards external objects, and consider the operation of causes, we are never able, in a single instance, to discover any power or necessary connexion; any quality, which binds the effect to the cause, and renders the one an infallible consequence of the other. We only find, that the one does actually, in fact, follows the other, the impulse of one billiard-ball is attended with motion in the second. This is the whole that appears to the *outward* senses.... From the first appearance of an object, we never can conjecture what effect will result from it. But were the power or energy of any cause discoverable by the mind, we could foresee the effect, even without experience; and might, at first, pronounce with certainty concerning it, by the mere dint of thought and reasoning. In reality, there is no part of matter, that does ever, by its sensible qualities, discover any power or energy, or give us ground to imagine, that it could produce any thing, to be followed by any other object, which we could denominate its effect.... It is impossible, therefore, that the idea of power can be derived from the contemplation of bodies, in single instances of their operation; because no bodies ever discover any power, which can be the original of this idea.
>
> (*EHU* 29–51 *passim*)

As we all know, it is easy to be swept along by Hume's elegant reasoning and to conclude, with Hume, that because nothing regarding power and causality is ever actually presented in our experiences, there must not be anything therein to discover. Power and causality, in other words, are chimeras, metaphysical analogues to the antipodean myths of the eighteenth century, because whenever we look for them, we fail to find them.

Thus, it is not until we pause and reflect on the methodological presuppositions of Hume's larger project that we find reasons to resist this seductive conclusion. Hume's project overly epistemologizes the case against the causality of physical objects and inherent causal powers. It is Hume's peculiar brand of empiricism, driven by his commitment to his Copy Principle, that underwrites this argument, as Hume himself acknowledges (*EHU*, 50). As is well known, Hume deployed the Copy Principle as a check on the significance (or meaningfulness) of terms with regard to their conceptual content: a necessary condition for the significance of a term was that the idea functioning as its meaning be 'copied from' a corresponding impression. This commitment to his Copy Principle severely restricts the nature of his empiricism, making it out to be almost verificationist. Both the 'copied from' relation and the restriction on impressions as relata constrain the conceptual content of the idea such that it must be manifest within one's immediate experiences, which is why there is such a push in Hume's argument toward the need to see or observe the causal power itself, the power as a power, force, or energy residing within the cause. This notion of manifestness, which underlies the restrictions, is quite powerful. Hume believed that any causal power would have to be evident per se, and that its evidentness would make any inference or explanation in support of it otiose; it should be clear to any open and inquiring mind, in other words, as it exists in its own right and by itself. It would therefore be perfectly obvious what the causal power was, that it could not be other than what it is, and that it acts or operates in precisely the way that it does independently from all other relationships that might intrude on it. There could, moreover, be no basis for questioning it or for seeking an explanation for it—the impression of it would carry all the evidentness the case would require and could demand. The situation would be analogous to the case of colour, in Hume's eyes: to demand an explanation of why red is, or appears, as red is absurd. Of course, red things might not be red, but impressions that appear red have to be red because that is precisely what being red means! To appear otherwise is to literally be otherwise; case closed, nothing more needs to be said or even could be said. Were power, force, or energy like this, it would analogously be perfectly obvious what it would be or do and analogously obvious that it could not be other than what it is or do other than what it does. Nothing like that is obvious in quite that way, however; you can look for yourself and know that. So, we are to conclude, nothing that constitutes a causal power is per se evident in the way that Hume's empiricism requires.

Hume's presentation of his argument against inherent causal powers in *A Treatise of Human Nature* makes the central role of this methodological Copy Principle even clearer than the above material from the *Enquiry*:

> I begin with observing that the terms of *efficacy, agency, power, force, energy, necessity, connexion,* and *productive quality*, are all nearly synonymous; and therefore 'tis an absurdity to employ any of them in defining the rest. By this observation we reject at once all the vulgar definitions, which philosophers have given of power and efficacy; and instead of searching for the idea in these definitions, must look for it in the impressions, from which it is originally deriv'd.... since reason can never give rise to the idea of efficacy, that idea must be deriv'd from experience, and from some particular instances of this efficacy, which make their passage into the mind by the common channels of sensation and reflection. Ideas always represent their objects or impressions; and *vice versa*, there are some objects necessary to give rise to every idea. If we pretend, therefore, to have any just idea of this efficacy, we must produce some instance, wherein the efficacy is plainly discoverable to the mind, and its operations obvious to our consciousness or sensation. By the refusal of this, we acknowledge, that the idea is impossible and imaginary;.... Our present business, then, must be to find some natural production, where the operation and efficacy of a cause can be clearly conceiv'd and comprehended by the mind, without any danger of obscurity or mistake.... [The conclusion that we have no adequate idea of power or efficacy from any object; since neither in body nor spirits, neither in superior nor inferior natures, are we able to discover one single instance of it] is unavoidable upon the hypothesis of those, who maintain the efficacy of second causes, and attribute a derivative, but a real power and energy to matter. For as they confess, that this energy lies not in any of the known qualities of matter, the difficulty still remains concerning the origin of its idea. If we have really an idea of power, we may attribute power to an unknown quality: But as 'tis impossible that the idea can be deriv'd from such a quality, and as there is nothing is known qualities, which can produce it; it follows that we deceive ourselves, when we imagine we are possest of any idea of this kind, after the manner we commonly understand it. All ideas are deriv'd from,

and represent impressions. We never have any impression, that contains any power or efficacy. We never therefore have any idea power.

(T, 1.3.13, 1:106–8)

We nowadays feel no such need to adopt Hume's restrictive Copy Principle—why should we? It is not news, and hardly a surprise, that following Hume's gambit forces us to lose half of our metaphysical world. But do we now have an answer to Hume? Maybe, but even if we don't, there is no reason why we have to follow his lead. Analytic philosophy returned long ago to a different form of empiricism, one which is decidedly more permissive and friendly to indirect observations of natures and essences. It is now easy, therefore, to dismiss Hume's own argument against causal powers as misguided and unsound. By avoiding Hume's peculiar epistemology, in other words, we can avoid his criticisms and return to the explanation he dismissed as 'more popular than philosophical': 'I believe the most general and most popular explication of this matter, is to say, that finding from experience, that there are several new productions in matter, such as the motions and variations of body, and concluding that there must somewhere be a power capable of producing them, we arrive at last by this reasoning at the idea of power and efficacy. But to be convinc'd that this explication is more popular than philosophical, we need but reflect on two very obvious principles' (T, 1:106). But how does that really help defend the postulation of causal powers? How does turning away from Hume's gambit and returning to the popular explanation of causal powers help metaphysicians ground causal powers and defend their postulation?[1]

Hume attributes the popular view to Locke (Essay, II.xxi.1, 233; see also II.xxi.4, 235) and describes it as follows: 'that finding from experience, that there are several new productions in matter, such as the motions and variations of body, and concluding that there must be somewhere a power capable of producing them, we arrive at last by this reasoning to the idea of power and efficacy' (T, 1:106). The precise nature of this argument is none too clear from the texts themselves. Locke wrote:

The Mind, being every day informed, by the Senses, of the alteration of those simple *Ideas*, it observes in things without; and taking notice how one comes to an end, and ceases to be, and another begins to exist, which was

---

[1] Others might want to focus on the two obvious principles that show the explanation to be more popular than philosophical, which Hume points to in the passage. That would be great, but I'm interested in going in another direction.

not before;…and concluding from what it has so constantly observed to have been, that the like Changes will for the future be made, in the same things, by like Agents, and by the likes ways, consider in one thing the possibility of making that change; and so comes by that *Idea* which we call *Power*.   (*Essay*, II.xxi.1, 233)[2]

The texts speak only of 'concluding from' the data of observation. It is quite likely that, as a historical point of fact, Locke meant this to be a kind of abductive argument—the best explanation for the observed phenomena of change is an inherent active power in the thing causing the change and a corresponding passive power in the thing being changed. But Locke does not quite come out and say that. Look closely at what he does say. Following up with the above passage from II.xxi.1, Locke explains that 'since whatever Change is observed, the Mind must collect a Power somewhere, able to make that Change, as well as a possibility in the thing it self to receive it' (*Essay*, II.xxi.4, 235). Conceptually, the actuality of change entails the possibility of change— Locke's text can be read as stating exactly that. But he was clearly aware of alternative accounts that recognize change and the possibility of change without the postulation of inherent powers in things. This alone suggests that his explicit invocation of powers was meant to be metaphysically significant— why else would he so prominently accept and display the language and concepts of powers unless he intended them to be read against the prominent alternative accounts? Since one of these alternative accounts was the doctrine of occasionalism, and in 1693 Locke engaged with a sustained discussion of the doctrine, we should consider what reasons he had for rejecting it. Thus maybe, if we turn there, we can more clearly see that his reason for accepting the reality of causal powers, as presented in this 'popular explanation', was largely abductive. Let's look at Locke against occasionalism, then.

Between late 1692 and mid-1693, Locke was considering adding a chapter to the *Essay* countering the Malebranche–Norris doctrine of seeing all things in God.[3] Part of the notes Locke drafted in support of this concerned their commitments to occasionalism as a component within their epistemological systems. Locke's general strategy against occasionalism is to flag the allegedly

---

[2] Locke's discussion of the idea of power has been subject to severe criticism by scholars, especially from a Humean perspective. For two treatments that offer a more sympathetic reading of Locke's discussion and its relationship to Hume's, see Coventry 2003 and Connolly 2017.

[3] The *Examination of P. Malebranche*, as it has come to be called, was not published until 1706 as part of the *Posthumous Works of Mr. John Locke*; the *Remarks on Norris*, as it has come to be called, was not published until 1720 as part of *A Collection of several Pieces of Mr. John Locke, Never before printed, or not extant in his Works*.

absurd consequences of the thesis, especially those that contravene the occasionalist's own core principles regarding divine creation. In effect, he argued that Malebranche's and Norris's commitments to occasionalism were internally inconsistent. The rub, according to Locke, was their commitment to the principle that God does nothing in vain (Norris) or that God acts in the simplest ways conceivable (Malebranche). If the thesis of occasionalism were true, Locke argued, all the wondrous complexity of the sense organs and the mediating structures of sense perception would be in vain: 'How will they reconcile to this principle of theirs [God does nothing in vain], on which their whole system is built the curious structure of the eye and the ear; not to mention the other parts of the body.... All that nice and curious structure of those organs is wholly in vain' (Locke 1720, 157). Locke himself certainly did not believe this to be 'lost labour', and he did not believe that the occasionalists really did either. He exclaimed that 'He that understands Opticks ever so little', as the occasionalist surely do, 'must needs admire the wonderful make of the eye, not only for the variety and neatness of the parts; but as suited to the nature of refraction, so as to paint the image of the object in the *retina*.' In Locke's eyes, then, occasionalism is ultimately driven by a fallacious argument from ignorance and the desire to be omniscient: 'But we do not know how any natural operation this can produce an Idea in the mind: and therefore (a good conclusion!) God the author of nature cannot this way produce it. As if it were impossible for the Almighty to produce any thing, but by ways we must conceive, and are able to comprehend: when he that is best satisfied by his omniscient understanding, and knows so well how God perceives and man thinks' (158). This is a common argument form for Locke. He was clearly hostile to the idea of rejecting something because we finite minds, in our fallen state of ignorance, could not conceive and explain how an omnipotent God might manage it.

But Locke does not leave matters merely at identifying the fallacy in the occasionalist's reasoning. He additionally provides the abductive argument for the existence of powers that was alluded to in II.xxi.1 and 4 (165–71), which is Locke's most sustained discussion of occasionalism. The argument begins with an affirmation typical of the early modern causal powers theorist: 'I think it cannot be deny'd that God having a power to produce Ideas in us, can give that power to another: or, to express it otherwise, make any Idea the effect of any operation on our bodys. This has no contradiction in it, and therefore is possible' (165). Then he identifies three natural phenomena that are not readily explained by the occasionalist hypothesis but are easily explained by the hypothesis of inherent causal powers. The first is the fact that objects are not occasional causes for blind people yet are for sighted people (166). The second is the aforementioned complex structure of the eye (166). The third is the phenomenon of sensory habituation, in which the occasional cause remains unabated yet the idea disappears from our consciousness (167). Then Locke draws out a canonical abductive inference. In response to the first phenomenon, he merely suggests that it 'wou'd give one reason to suspect something more than a presential *occasional cause* in the object' (166). This suspicion is then vindicated by the next two phenomena: "Tis hard to account for either of these Phaenomena, by God's producing these Ideas upon the account of *occasional causes*. But *by the production of Ideas in the Mind, by the operation of the object on the organs of sense; this difference is easy to be explained*' (167).

Locke talks in this argument of the 'operation of the object'. How does a commitment to the existence of causal powers come out of that? We should note that Locke's depiction of this 'operation of the object' is again wholly couched in the terminology of powers in what immediately follows. Locke added:

> For I hope they will not deny God the privilege to give such a power to motion, if he pleases .... The infinite eternal God is certainly the cause of all things, the fountain of all being, and power. But because all being was from him, can there be nothing but God himself? or because all power was originally in him, can there be nothing of it communicated to his creatures? This is to set very narrow bounds to the power of God, and by pretending to extend it, takes it away. For which (I beseech you, as we can comprehend) is the perfectest power; to make a machine, a watch for example, that when the watch-maker has withdrawn his hands, shall go and strike by the fit contrivance of the parts; or else requires that whenever the hand by pointing to the hour, minds him of it; he should strike twelve upon the bell? No machine of God's making can go of it self. Why? because creatures have no power; can neither move themselves, nor any thing else.   (Locke 1720, 169)

There is a lot here to suggest that Locke was advocating for an abductive argument in support of the postulation of causal powers.[4] It looks like Locke is pushing a position very much like the doctrine of 'structure realism' and

---

[4] Locke's qualms about mechanistic reductions have been long discussed by scholars (Wilson 1979; Ayers 1981; Wilson 1982; Downing 1998; Stuart 1998; Rozemond and Yaffe 2004; Connolly 2015; Wood 2016), and they fit perfectly with the emphasis on powers in a broadly Aristotelian sense throughout the *Essay* as well as with the attacks on occasionalism.

'structural powers' currently being pushed by neo-Aristotelians (Simpson et al. 2018). At least, this is not an implausible way of reading the texts and Locke's appeals to the language and concepts of causal powers. This, therefore, makes the contemporary causal powers debate look less like a division between the scholastics and the moderns and more like a division that cuts across both camps and tracks a division between causal powers realists, on the one hand, and occasionalists and sceptics on the other.

Unfortunately, Locke ignores the worries and concerns that are pushing the occasionalists and the underlying metaphysical basis of their system. And this is what's interesting and significant for the contemporary discussion, I think; not what Locke says, but what he doesn't say and why he doesn't say it. It is not sufficient to declaim against the argument from ignorance and to simply postulate the existence of causal powers, as Locke does. It is not sufficient because that answers a different question than the one raised by the occasionalists. Locke, and others who appeal to an abductive argument, answer the question of whether or not causal powers obtain. But the question pressed by the occasionalists' worries and concerns is not whether they obtain, but how do or might they obtain. This suggests the need for an explanation of causality and power rather than an argument supporting the existence or postulation of causal powers. That subtle difference in context is forced by their metaphysical worries about the natures of causal powers and substance and significantly alters grounds for a successful or proper answer. Locke misses that difference by paying attention only (or primarily) to the thesis itself and not the underlying motivations that animate the occasionalists' thought processes. What's at issue in the seventeenth-century debate is not whether causal powers exist, but how they could, given the understanding of substance they were struggling with.

And that's the real problem that I'd like to suggest contemporary causal powers theorists ought to focus on. To set up causal powers as ontological primitives in the construction of an abstract system misses the essential philosophical point, which is uncovering and exploring some sort of explanation for how causal powers can subsist and perform their functions as causal powers. Indeed, that seems to magnify the worry and its problematic character. 'Biting the bullet' or simply accepting them as a 'starting point' are merely dialectical moves sidestepping the philosophically interesting and significant question. That's all well and good if one wants to play that intellectual game, but Hume would be the first to emphasize that it is a merely intellectual game, I think. The worries and concerns underwriting the occasionalists' question remain untouched by it, which is why it continues to intellectually rub and fester like a too-small, wet leather shoe on a forced march. We can all acknowledge that the occasionalists' reaction to this rubbing, which is to give up on the possibility of explaining how causal powers are possible and pushing all efficient causality up to divine agency, is undesirable. We can also all acknowledge that Hume's resolution, emphasizing that we don't need any explanation because we don't need causal powers because mere regularities will do, is equally undesirable. But neither resolves the underlying worries. The former is like suggesting that we just change to a different pair of shoes (which may be equally tight, wet, and chafing, but seem acceptable because they rub in a different spot), and the latter is like suggesting that we don't need shoes and can just march barefoot. Better would be to look into where, how, and why the shoe rubs and work to repair it. But certainly, telling ourselves, and our compatriots in identical positions, that the shoe doesn't really rub or that we have to continue wearing it, and so shouldn't really complain about it because it is the only shoe we have, is not a proper solution worthy of a master shoemaker. We'd expect a shoemaker to fix it or to offer an explanation of why no repair is needed or is possible. Hume and the occasionalists are holding us metaphysicians to the same standards, and rightly so.

We therefore ought to turn our attention to the occasionalists' worries and concerns and try to diagnose what causes them to rub so that we can provide the requisite explanation or explain why such an explanation would be unnecessary or unattainable.

2. Seventeenth-century Occasionalism

In the seventeenth century, occasionalism was an almost exclusively Cartesian phenomenon. Not all Cartesians were occasionalists; but almost every occasionalist was considered to be a Cartesian or strongly influenced by Descartes. Nicolas Malebranche was the most famous of them, but their numbers included Frenchmen Louis de La Forge and Géraud de Cordemoy, Fleming Arnold Geulincx, and Englishman John Norris. There seemed to have been something about Cartesianism that led to the development and defence of occasionalism (Nadler 2011). It would be a mistake, however, to infer that the underlying worries and motivations animating the occasionalists were exclusively Cartesian or that we now easily avoid occasionalism and its underlying motivations because we are not Cartesians. To be sure, many of their arguments for occasionalism rested on distinctively Cartesian premises and were explicitly directed at other Cartesian thinkers. But that's a different matter than the intuitive core of the underlying worries and motivations, which is what we are focusing on. The lesson for us lies in those underlying intuitive motivations and worries. But first, a bit of background.

Scholars commonly identify four central arguments that were made on behalf of occasionalism (Lee 2018; 2019): the Passive Nature Argument, which rests on the claim that body is passive by nature and incapable of initiating motion; the No Knowledge Argument, which rests on a requirement that causality requires a causal agent with knowledge of the means to achieve the effect; the No Necessary Connection Argument, which rests on the requirement that causality requires a necessary connection and the claim that causal necessity is no genuine type of necessity; and the Conservation is but Continuous Creation Argument, which rests on the claim that the conservation of substances is a form of continuous divine recreation at every moment of their existence. From our perspective, as contemporary practitioners interested in learning from the past, the arguments used by the Cartesians and directed at other Cartesians are not the main object of interest. They are not irrelevant, of course; but we need to shift our gaze a bit. None of us are Cartesians, after all, and we're not the target audience for the arguments. As stated above, it's what made the arguments powerful and seductive to the Cartesians that is of interest to us. So, when returning to the historical texts, we want to focus more on what the Cartesians say about the arguments than their formulations of the arguments themselves. We need, in other words, to look at the context surrounding their presentations of the arguments to learn the lessons from the past, which we are interested in.

I'd like to draw our attention to three underlying worries, which seem to me to be most relevant to our attempt to revive the Aristotelian notion of causal powers. These are three concerns that seem most damaging to the idea of causal powers in the eyes of the Cartesian occasionalists and seem to be most likely to menace the current movement. The first concerns the conceptual clarity of the metaphysical category corresponding to causal power. In seventeenth-century jargon, this would be the clarity of the idea of power or efficacy, but what they had in mind was a particular sort of obscurity contained within the idea—what kind of metaphysical entity is a power and how does it connect with its subject? The second worry concerns the idea of fully determining the character of the effect. Causal-power talk is generally focused on type–type relationships and categories of phenomena, but all actual causal events are non-idealized particulars with precisely determined outcomes that are generally diverse from one another, even when occurring in essentially identical contexts. Seventeenth-century occasionalists struggled to explain how the same causal power in virtually identical situations produces slightly different effects. This struggle pushed them to conceptualize causality in terms of agency, which we will explore in more detail below. The third worry concerns the nature and grounds for natural necessity. They were worried about how to explain why certain relations were naturally necessary when others weren't, and what natural necessity consisted in.

Malebranche is perhaps the best figure to focus on because he was the most sophisticated thinker of the bunch and his work effectively synthesized the thoughts and concerns of most other occasionalists. But I will draw upon the works of others as appropriate. Malebranche's primary occasionalist works are *The Search After Truth* (1674–5), the *Elucidations of the Search after Truth* (1678), and *The Dialogues on Metaphysics and Religion* (1688). These will serve as the core of my discussion.

2.1. The No-conceptual-clarity Worry

When Malebranche opens his discussion of occasionalism, he calls the doctrine of inherent causal powers 'the most dangerous error of the philosophy of the ancients' (*Search* 6.2.3, 446). That's weird to our ears, but for Malebranche it's dangerous because of the anti-Augustinian conclusions regarding human nature and the lack of intimacy in our relationship to God. Thus, believing that physical objects have causal powers might lead us to 'render sovereign honor to leeks and onions', for example. Weird, but it makes sense in the *Dialogues on Metaphysics and on Religion*, where the topic is philosophical theology. However, in the *Search*, which is a work in logic (in the seventeenth-century sense), a discussion of the dangers of believing in inherent causal powers seems out of place.

Malebranche brings it up, however, as an example of the ancient philosophers' failure to follow the proper logical method. When reading Malebranche's discussion, the emphasis needs to be placed on the erroneousness of the belief in causal powers, an erroneousness due to poor epistemic practices. The primary worry motivating Malebranche is expressed in the following passage:

> If we assume, in accordance with their opinion, that bodies have certain entities distinct from matter in them, then, having no distinct idea of these entities, we can easily imagine that they are the true or major causes of the effects we see. That is even the general opinion of ordinary philosophers; for

it is mainly to explain these effects that they think there are substantial forms, real qualities, and other similar entities. If [,however,] we next consider attentively our idea of cause or of power to act, we cannot doubt that this idea represents something divine.   (*Search* 6.2.3, 446)

The error consists in believing where we lack a distinct idea of 'these entities' and imagining that they are the true or major causes of the phenomena of change. Just as we would expect, in the Elucidations this focus on the lack of a clear and distinct idea of causal powers is brought even further into the limelight. Malebranche sees the continued commitment to inherent causal powers as a kind of prejudice caused by the Fall and our embodiment, and clarifies his grounds for opposing it in hopes that it will help our reason to triumph over the sensory inclinations and the imagination. He explains:

> There are many reasons preventing me from attributing to secondary or natural causes a force, a power, an efficacy to produce anything. But the principal one is that this opinion does not even seem conceivable to me. Whatever effort I make in order to understand it, I cannot find in me any idea representing to me what might be the force or power they attribute to creatures. And I do not even think it a temerarious judgment to assert that those who maintain that creatures have a force and power in themselves advance what they do not clearly conceive. For in short, if philosophers clearly conceived that secondary causes have a true force to act and produce things like them, then being a man as much as they and participating like them in sovereign Reason, I should clearly be able to discover the idea that represents this force to them. But whatever effort of mind I make, I can find force, efficacy, or power only in the will of the infinitely perfect Being.
>
> (*Search*, Elucidation Fifteen, 658)

This is, obviously, the analogue to Hume's suggestion that we have no idea of power and efficacy because experience fails to provide us with any. In Malebranche, of course, the suggestion is not bound up with Hume's narrow empiricism. Malebranche's suggestion is importantly different in a couple of ways. First, Malebranche's is entirely conceptual. That our idea of natural power and efficacy has no genuine cognitive content is what Malebranche is claiming. Second, the lack of content revolves around the metaphysical character of power and efficacy. This is brought out in the very next paragraph when he criticizes the diversity and lack of consensus among the Aristotelians regarding the nature of power. They appeal to 'substantial forms', 'accidents or qualities', 'matter and form', 'form and accidents', and 'certain virtues or faculties', Malebranche argues, and these are not just different names for the same thing because they all take different stands on how causality operates; 'some maintain that the substantial form produces forms and the accident form accidents, others that forms produce other forms and accidents, and others, finally, that accidents alone are capable of producing accidents and even forms' (*Search*, Elucidation Fifteen, 658). The appeal to a lack of consensus is interesting here. It's not an ad populum fallacy, but rather is being presented as evidence of the conceptual vacuousness of the idea itself. Such a diversity of speculations obtains precisely because no one knows what they are talking about, since there are no grounds to guide our metaphysical judgements regarding causality. That the diversity and disagreements concern the forms and the relationships between forms and powers strongly emphasizes this. What's at issue, then, beneath Malebranche's denial are not correlations between causes and effects, or even the details of causal processes, but the metaphysical structures that constitute causality.

Even though Malebranche's claim is couched in the Cartesian terms of clear and distinct ideas, one does not really need to be a Cartesian to feel the underlying intuitive pull of his point. It does seem very difficult, upon thoughtful reflection, to say that the community of philosophers and metaphysicians has made much progress to identifying and analysing the metaphysical features involving causality. Indeed, this very point seems to be the take-home message of the Stanford Encyclopedia of Philosophy entry on Dispositions (Choi and Fara 2018): lots of work on the semantics of how we think and talk about causal dispositions has been done, but there is no real consensus about even that and very little about the underlying metaphysical structures and nature of substances and properties that would properly constitute an answer to the question, 'what are powers?' If we take the SEP as representative of the current state of philosophical understanding, as seems reasonable, it is hard to see our current state as essentially much different than that of Malebranche's, despite the different ways of expressing what sort of relationship we would recognize as a causal one. And that's a real worry. This might not seem so surprising if we remember the original reason why the concept of powers was introduced: Aristotle introduced potentiality as a bridge between being and non-being in a move to avoid the Eleatic paradox of becoming. The root concept is, thus, a rather bastardized notion combining elements of both being and non-being—a power is a quasi-existential state such that a property does not actually obtain but would obtain were the power actualized (to differentiate it from an absolute state of non-being) and when actualized it in fact obtains and is transformed into a fully existential state. There is nothing conclusive that says such a bastardized notion is impossible or inconsistent; yet there is also nothing conclusive that says that it is isn't. Hence the lack of consensus. In Aristotle's day, it might have been dialectically acceptable to accept it and hope for, or even expect, greater clarity to eventually come. By the time of the Cartesian occasionalists, however, there had been over 2000 years for clarity to be achieved, and not even some of the greatest metaphysicians who had ever lived were able to make much progress, if any. That's the real source of the underlying intuitive pull of Malebranche's claim.

It might seem as if the lesson to learn is not to reintroduce causal powers until we finally resolve what their metaphysical features are, and some might object that this is too high a standard to hold contemporary metaphysicians to. But if we are not going to ignore and overlook this lesson of the seventeenth-century occasionalists, caution toward accepting causal powers into our metaphysics until the concept can be clarified is certainly warranted. Even if demanding that the metaphysical questions be clearly resolved is too strong, the dialectical state does demand that we should perceive something different this time around before giving causal powers a second chance at taking a leading role in our metaphysics of body. Clear and compelling accounts of their metaphysical features would be sufficient, but not necessary. What is demanded of proponents of causal powers instead is evidence that conceptual advances since the seventeenth century give us new, promising avenues and resources for resolving the metaphysical mystery of what causal powers are. Evidence of that does not seem to be too much to demand. It is certainly the case that there is considerable enthusiasm, and work is now being done again on the metaphysics of causal powers and efficacy. And such work should be encouraged, supported, and championed, to be sure. That is a different matter from advocating the return of causal powers to our metaphysics or philosophy of science. But I don't think that we can really say that such work has yet provided evidence that the mystery of what causal powers are is likely to be cleared up. And that's what the dialectical state would seem to require before we build Aristotelian causal powers back into our metaphysics of bodies and our philosophy of science. Malebranche emphasizes this in the *Dialogues*: 'It is up to you, Aristes, to give me some idea of this power....It is better to remain silent than not to know what one is saying' (Malebranche 1997, dial. 7, 107). Perhaps there is such evidence, or a case can be made on behalf of such advances—Great, let's see it! But let's see that rather than abductive arguments merely maintaining that causal powers ought to be accepted or returned to the centre of our metaphysics or our philosophy of science.

Malebranche pushes yet another feature of the debate that seems to magnify the importance of this dialectical point that the onus is on the causal-powers theorists to show progress in clarifying the concept of a power. He flags the commitment to inherent causal powers as a prejudice instilled in us by repeated experience. While I would not wish to suggest that prejudice plays any role in contemporary philosophical discussions, Malebranche's appeal to repeated experience grounding belief is reminiscent of Locke's popular argument—experience shows us that causes act to produce effects and the best explanation for this is that bodies have inherent causal powers acting and passively receiving those actions. This helps to drive the point home that admitting causal powers cannot rest on what seems obvious or natural to us. We need to make sure that their acceptance is properly grounded and that we are not simply falling for a natural, common-sense presumption when accepting their existence on the basis of Locke's popular argument. The dialectical state in the seventeenth century as well as now, in other words, means that we really need clarity regarding the metaphysical features or evidence that such clarity is likely to be developed, before we ought to return to a metaphysical commitment to causal powers. This suggests a slightly different direction for the contemporary discussion.

2.2. The Explaining-full-determination Worry

There is another feature of the occasionalists' critiques of secondary causes that is suggestive of an interesting worry about the alleged operations of causal powers. I find this worry to be especially interesting because it lies at the intersection of two very different arguments for occasionalism, the so-called No Knowledge Argument and the No Transference Argument. The No Knowledge Argument might be familiar because, thanks to Malebranche, it is commonly identified as a main argument for occasionalism. But the No Transference Argument might not be familiar because it was presented by the obscure Cartesian Louis de La Forge and presented only in support of a restricted version of occasionalism.

The No Knowledge Argument sounds very odd to our ears now:

For how could we move our arms? To move them, it is necessary to have animal spirits, to send them through certain nerves toward certain muscles in order to inflate and contract them, for it is thus that the arm attached to them is moved....Men who do not know that they have [animal] spirits, nerves, and muscles move their arms, and even move them with more skill and ease than those who know anatomy best. Therefore, men will to move their arms, and only God is able and knows how to move them.

(*Search* 6.2.3, 449–50; cf. *Search*, Eludication Fifteen, 670–1)

Why would Malebranche equate causality with agency? It's a weird assumption and makes it easy to summarily dismiss the argument as irrelevant for discussions of natural causation. What could possibly be the motivation for accepting such a presumption?

Perhaps it is driven by a commitment to the principle that if one wills the end, one must will the means. To will the movements, we must will the complex chain of constituent movements that constitute or make up that movement. In Malebranche's example, in willing to move our arm, we would have to also will to activate the right neurons in just the right sequence and to exactly the right degree to actually move our arm in the appropriate manner. Since willings have to be direct, perspicuous, and informationally precise, any proper volition to move our arm would have to involve a host of volitional subcomponents, each of which we'd have to also be conscious of. Of course, there's a lot here that a non-Cartesian would object to and could avoid in order to preserve a commitment to this principle. But really this all misses the central point—why equate causality with agency? Even if we understood agency better than natural causality, it does not at all seem appropriate to use agency as an analogue in analysing natural causality.

If we look at the texts supporting and developing the arguments from the *Search*, however, I think that we can discern an underlying worry that the move to equating natural causality with agency is supposed to solve. In the *Dialogues*, Theodore guides Aristes from the neutral position of neither affirming nor denying that bodies can act on other bodies with the following line of reasoning: 'Do you not clearly see that bodies can be moved, but they cannot move themselves? You hesitate. Well, suppose then that this chair can move itself. In which direction will it go, with what speed, and when will it decide to move itself? Give it then an intelligence as well, and a will capable of determining itself. In other words, create a human being out of your armchair. Otherwise, this power of self-motion will be useless to it' (Malebranche 1997, dial. 7, 110–11). The text suggests that the move to equating natural causality with agency is the solution to the problem of full determination. When bodies move, undergo change, or develop, they develop in fully determinate ways. Existence is fully determinate, of course. So, when something changes from rest to motion, it must move in a fully determinate way—fully determinate with regard to direction, velocity, spin/gait/character, and all concomitant features of moving. Aristes's armchair, for example, were it a self-mover, would begin moving at a definite point in time, in a certain direction, at a certain speed, for a certain distance, with a certain gait or type of movement, etc. What determines all that; what could determine all that? This is the problem of full determination: how do we explain the full determination of this instance of the activation of the power? If it were a self-mover, it would have to 'choose', either freely or determinately—hence Theodore's strange suggestion to 'create a human being of your armchair'. Aristes is having none of that, of course: 'A human being out of my armchair! What a strange thought!...I certainly think that this chair cannot move itself' (111).

But the problem of full determination arises even if the object is put into motion by something else, God included. Malebranche, as Theodore, continues, arguing, 'For even God—though omnipotent—cannot create a body which is nowhere or which does not have certain relations of distance to other bodies. Every body is at rest when it has the same relation of distance to other bodies; and it is in motion when this relation constantly changes. Yet it is obvious that every body either changes or does not change its relation of distance. There is no middle ground' (111). Imagine being an animator on Disney's *Beauty and the Beast*—the armchairs can skitter, gallop, skip, slide, etc. across the floor; how do you want to handle their movements? The domain of possibility is virtually endless, yet the movement has to be fully determinate, a single expression from this virtually infinite domain. Malebranche conceptualized the solution in terms of agency—the chair chooses, or someone else chooses, or God chooses; not the chair; not someone else; therefore.... Some form of agency is the solution to the problem of full determination and grounds explanations for instances of motion or change. But at the cost of foisting large epistemic requirements on the active object. The epistemic consequences are not our concern, however (the plausibility and grounds for occasionalism itself are not our issue). But this underlying worry regarding the problem of full determination is our concern—what fully determines the various characteristics in the activation of a power?

Nowadays we are inclined to consider the complexity of the context as imposing itself on the activation of a power. In other words, we tend to think that the reason why different activations of the same determinable power display different determinate properties is because the actors and their context differ subtly and in a myriad of ways; it has nothing to do with choice from among a virtually infinite domain of possibilities but from the virtually infinite complexities among the natural objects constraining the exemplification of the power. Malebranche, in other words, was rather naïve in believing that such complexity required an agent to oversee and manage it.

Fine, fine: I have absolutely no doubt that the recent advances in mathematical modelling and computational power necessary to work through such models will give us much clearer and richer understandings of the complexities actually involved in the exemplification of any causal power. Think of modelling the chemical and histological complexities involved in the short-term and long-term impacts of a can of Coke on the human liver, for example. But the point for us as philosophers is, what does this sort of response do to the metaphysics of neo-Aristotelian hylomorphism? That is the challenge that the problem of full determination poses for contemporary proponents of causal powers. By focusing on the complexities of the interactions and characteristics of the actors involved and the context in which they are interacting, the subject–power relationship, central to the Aristotelian metaphysical system, is threatened in that the power seems to be an emergent property of a state of affairs rather than a per se property of a singular substance. Because determinable powers are conceptually tied to the possibilities of their determinate instantiates, the more that instantiations are relativized to fully determinate contexts and interactions, the less coherent the category of causal powers becomes, and the less explanatory power any such appeals to them carry.

This horn of the dilemma can be avoided by connecting the full determination of a power's activation to the force or energy necessary to activate the power rather than the complex interactions and context involved in its exemplification—diverse determinations of an effect in seemingly identical contexts are thus explained by different levels, or different types of energies that activate the causal power. But this has metaphysical baggage of its own that also threatens neo-Aristotelian hylomorphism. In this case, the problems revolve around the ontological status of this energy or force and its relationship to the substance of the actors involved in the exemplification of the power. Louis de La Forge's No Transfer Argument is a nice analogue to this worry.

La Forge used the argument to argue against the possibility of mechanistic body–body causation, causation involving the transference of motion from one object to another, and set up the occasionalistic alternative:

> We should distinguish between a movement and its determination and between the cause of a movement and the cause which determines it, because one is often different from the other just as movement is from the

force which makes something move.... Motion is only a mode which is not distinct from the body to which it belongs and which can no more pass from one subject to another than the other modes of matter.... But the motive force, i.e. the force which transports a body from one vicinity to another and which applies it successively to different parts of the bodies which it leaves behind, which is also sometimes called 'motion', is not only distinct from this application but also from the body which it applies and moves.... Now if the force which moves is distinct from the thing which is moved and if bodies alone can be moved, it follows clearly that no body can have the power of self-movement in itself. For if that were the case this force would not be distinct from body, because no attribute or property is distinct from the thing to which it belongs. If a body cannot move itself, it is obvious in my opinion that it cannot move another body. Therefore every body which is in motion must be pushed by something which is not itself a body and which is completely distinct from it.

I may be told that I assume without argument that the force which moves must be distinct from the thing which is moved. But it is easy for me to show that, because if the force which transfers and thereby applies bodies to each other could belong to them in such a way that the thing which is moved were itself the principle of its motion and this force were identical with it, then the notion of this force would have to include in its concept the idea of extension, as the other modes of body do. This is not the case.... Let us assume, if you wish, that this force is a mode of body; it could not then be distinguished from it and consequently it could not pass from one body to another. If you conceive it in the same way as real qualities are conceived, in the Schools, and if you think it is definitely an accident of body, even if it distinct from it, then you would have to conceive that it subdivides itself when one body moves another and that it gives part of its movement to the other body, and is therefore itself a body, at the same time that you assume it is distinct from corporeal nature; for anything which is divisible and which has parts which can exist independently is a body; or you would have to say it does not subdivide but that the body in which it is present produces a similar property in the body it touches when it pushes it. You thereby give to bodies the power of creation, for if the motive force is distinct from body, it is a genuine substance despite the fact that you call it an accident whose being (if I may use the term), since it is not drawn from any other substance by division, can be produced only by creation. Besides, what happens to the first moving force when there is no longer any motion in the body which was moved? Will you say that it is annihilated?   (*THM* 145–6)

La Forge's argument is couched in the Cartesian ontological terms of 'substance' and 'mode' and is directed toward 'motive force', but we can replace 'motive force' with 'energy' and think in terms of properties or states rather than modes to formulate a similar sort of dilemma. The central ideas concern what energy is and how it connects with the substance that it actualizes or activates.

It is easy for us to talk about energy and the activation of powers in terms of 'energizing' them or of the transference of energy to them. But what is this energy that we are speaking of? La Forge's dilemma emphasizes that we have two ways of conceiving it: either energy is a special kind of substance or it is a property or state of a substance.

If the former, it has the requisite existential independence to be transferred from one body to another. In itself, that's fine, but what does it do to a neo-Aristotelian hylomorphic concept of substance? And, more importantly, how does it interact with the other body such that it activates its causal power? I have had difficulty locating sustained contemporary explanations of this. The traditional answer for what happens when one substance ingests another is that some sort of 'natural transubstantiation' occurs. Maybe contemporary practitioners would like to go this route, though I do not see why; likening the process of the absorption of energy to transubstantiation does little to naturalistically explain and ground the process. Maybe contemporary practitioners would like to say that transferred energy is merely co-located with a body and activates its causal powers on the basis of that co-location. That's again fine in itself. But it does not seem keeping in the spirit of an Aristotelian substance. For now, there is some sort of special, co-located stuff that is not strictly speaking part of the body yet is triggering the body's powers through its proximity. Either a regress looms (in that co-location is not sufficient to account for the activation of a power, but something from the independent substance, the energy stuff, must still be transferred to the body to trigger its causal power) or the hylomorphic structure of individual substance threatens to collapse (in that the body is composed of matter, form, and some sort of co-present energy). Some explanation of how this is consistent with hylomorphic principles seems to be in order. Furthermore, there is presumably some sort of natural, internal process co-related with this transference of energy and activation of a causal power. Whatever that might be, why not mechanistically conceive of the energy-body process and reduce the activation and exemplification of the causal power to the mechanistic process itself? In other words, if energy is conceived as an independent substance transferred into a body that is not a hylomorphic unity but rather an atomistic composite, what is the real metaphysical import of returning to Aristotelian causal powers? Some further explanation is needed regarding that, it seems.

An alternative would be to conceive of energy as a state or property of a body that is triggered by either something internal or something external to it. Again, that is fine in itself, but the consequences for the metaphysical framework of causal powers needs to be developed to accommodate that move. And it is not obvious that it can be. On this model, there is no genuine transfer of anything, but changes are happening nonetheless. Whether the source of the energy is internal or external, matter in a certain state is given or accorded the creative power to put matter into a different state through its proximity alone. La Forge does not seem to be misrepresenting the dynamics of this model for conceiving of the activation of a causal power when he likens it to creation ex nihilo or some sort of magical transformation from bodily state α to bodily state β without anything being transferred into the body. La Forge pushes us to demand some sort of explanation for this rather than to defer it. And so long as the states or properties of a body are characterized as 'that body's' (or 'that body's time-slice'), it is difficult to see how the neo-Aristotelian will provide such an explanation. The situation seems quite similar to the paradoxes involving substantial forms when a substance undergoes substantial change that Aristotle outlined in On Generation and Corruption.

A complete, detailed account of what goes on metaphysically within or between substances when a causal power is activated and a determinate effect produced certainly seems warranted, as La Forge and Malebranche would emphasize. And it is important that it at least addresses, if not resolves, these paradoxes involving agency and the transference of energy in ways consistent with the hylomorphic structure of substances and powers.

2.3. The Grounds-of-necessity Worry

It might seem easy to dismiss concerns about the coherence of natural necessity as Humean sceptical theatrics. But Malebranche was no sceptic and he shared Hume's underlying concerns about natural necessity. 'A true cause as I understand it is one such that the mind perceives a necessary connection between it and its effects. Now the mind perceives a necessary connection only between the will of an infinitely perfect being and its effects. Therefore, it is only God who is the true cause and who truly has the power to move bodies' (Search 6.2.3, 450).

The key for us to recognize is that this was not just hollow piety; it was also a central feature of Malebranche's response to a worry that is essentially the same as our worry about the coherence of natural necessity. The crux of the metaphysical problem is that causal powers are contingently necessitating features of our logical possible world. It's not just whether or not a causal power obtains within a world; it's also whether or not the powerfulness of the causal power only ever contingently obtains. For example, whether a piece of stemware is actually fragile depends on the existence of other things in its world solid enough to shatter it (or so the standard example goes), and vice versa for the hammer's having the power to shatter stemware. But there is still another issue in that the supposedly essential relationship between the universal properties of being fragile and being able to shatter glass is contingent as well, at least as far as we can see. Yet these contingent properties are supposed to be necessitating, naturally necessitating to be sure but necessitating nonetheless. How is that category metaphysically coherent?

Historical Aristotelians faced the same problem, and during the medieval and early modern periods conceptualized it through divine creation. They centred their thinking around explaining miracles and counterfactuals regarding what God could or could not do, but the core issue was the same as ours—if causal powers are merely contingent, how can they be, properly speaking, necessitating? The medieval and early modern discussions fell along a continuum between two extremes, the intellectualist pole and the voluntarist pole. The intellectualists grounded natural necessity in the Forms or the Divine Ideas themselves. This allowed them to be truly necessitating, but it put pressure on the contingency of the powers. The intellectualists' version did not allow God to be omnipotent in a strong sense, did not allow miracles, properly speaking, to obtain, and did not permit powers to be accidental features of the world. On their view, even though God might not create hammers (or stemware), any world that contains hammers (or stemware) is a world in which fragility and the power to shatter must obtain.

The voluntarists, on the other hand, left natural necessity ungrounded. For them, God made the powers powerful through his unrestricted act of creation. As a matter of fact, God made glass fragile and gave hammers the active power to shatter, and so for us (or relative to us) these powers are essential to stemware and hammers, but God did not have to create them such and neither follows from the other. Moreover, God could have made them different, or have made the metallic hammer out to be fragile and glass to have the active power to shatter. And God could have done this within our own logical possible world without changing anything else within or about it. Obviously,

God is omnipotent in a strong sense for the voluntarists, the metaphysical possibility of miracles is easy to explain, and powers, like everything else, are a contingent feature of our logically possible world.

These are the extremes. But after the voluntarists' critiques of the intellectualists' standard position appeared on the scene, intellectualists developed a weaker, compromise position. The essential move in the compromise position was simple—let God voluntaristically create the Divine Ideas, but then they become fixed and necessitating, even relative to God, because of God's own nature. This seems to allow for the counterfactuals necessary for omnipotence in a strong sense, the possibility of genuine miracles, and as much stability as nomologicality requires, as well as a proper grounding of the powerfulness of powers.

For Malebranche, this sort of move allows the imagination to fly free and chase down all sorts of counterfactual scenarios, allowing us to think about alternative ways God might create a world. But when we intellectually apprehend the Divine Ideas, we can now see the nomic necessity and appreciate God's wisdom for making just these actual causal connections. Non-occasionalists could avail themselves of something like this just as well as occasionalists can. The key is not that Malebranche and the occasionalists made God out to be the only efficient cause but is rather the balance between forms or natures being contingently structured yet necessitating. It is this balance that provides a proper characterization of natural necessity, should it obtain of course. For the early moderns, whether Aristotelian or occasionalist, because all the conceptual work is fobbed off onto God, the worries that causal connections are merely human conceits or accidental features of our limited experiences and conceptions are completely avoided.

Unfortunately, we don't have recourse to God to fob the responsibility and the conceptual work off onto. Contemporary metaphysicians need to discover a naturalistic way of finding such a balance between contingency and necessity, and then find it within our logically possible world. I confess, I have no idea what that might be or look like, or even where to begin characterizing it. Alan Sidelle has discussed the semantic and logical challenges that the contemporary essentialist conceptual framework faces in making sense of the contingently necessitating aspect of causal powers (Sidelle 2002). One of the advantages, perhaps, of approaching this through Malebranche's concept that only the divine will is necessitating is that we can see a reason why we have such trouble understanding the contingently necessitating character of causal powers—the origins and source of a power's powerfulness are opaque. If we understood the source for a power's powerfulness and could explain why there was such a relationship between active power φ and passive power π, perhaps we could grasp the reasons for its being necessitating even though it is contingent. The moderate voluntarist position provides such an understanding precisely because we can see why the relationship is necessary relative to our logically possible world (God willed it to be so), while also seeing why it is ultimately contingent (God could have willed otherwise). But from within a naturalistic framework, it is difficult to see what that source might have been, *a fortiori* if scientific theories entailing that our universe is one multiverse among billions upon billions of others.

3. Conclusion

The dominant themes of this romp through seventeenth-century occasionalist thinking are twofold. First, the dialectical state after the occasionalists' criticisms mean that we cannot just postulate causal powers as explanatory devices and expect or hope that we'll be able to eventually resolve any metaphysical issues that arise. The occasionalists took us past that point and we have an obligation to address their concerns, or at least provide reasons suggesting that this time is different, and we can be confident that their concerns will soon be resolved. Second, the core issue should not be thought to be whether or not causal powers exist or obtain; rather, the core issue is explaining how causal powers could be. The occasionalists' concerns, in other words, focus on what causal powers are, what is the nature of a causal power. Thus, the opacity of what a causal power conceptually consists in is pushed to the forefront. Abductive arguments and the need for them to support the truth conditions of counterfactual claims are thus irrelevant. The occasionalists insist on having explanations for how such bastardized entities could obtain, how they could be productive of phenomena, and how they could bind subsets of 'close' logically possible worlds such that they provide counterfactuals with truth conditions. This all means that, in our discussions about causal powers, they can no longer be presented as primitive or as metaphysical starting points. It is heartening to see a number of philosophers working in this direction and moving our concept of a causal power forward. It is hoped that this look back at occasionalism can encourage and support such efforts by bringing some of the underlying concerns that brought down Aristotelian causal powers in the first place into sharper focus.

References

Ayers, Michael. 1981. 'Mechanism, Superaddition, and the Proof of God's Existence in Locke's *Essay*'. *Philosophical Review* 90 (April): 210–51.

Choi, Sungho, and Michael Fara. 2018. 'Dispositions'. In *Stanford Encyclopedia of Philosophy*, rev. 22 June 2018. https://plato.stanford.edu/entries/dispositions/.

Connolly, Patrick. 2015. 'Lockean Superaddition and Lockean Humility'. *Studies in History and Philosophy of Science* 51 (June): 53–61.

Connolly, Patrick. 2017. 'The Idea of Power and Locke's Taxonomy of Ideas'. *Australasian Journal of Philosophy* 95, no. 1: 1–16.

Coventry, Angela. 2003. 'Locke, Hume, and the Idea of Causal Power'. *Locke Studies* 3: 93–111.

Downing, Lisa. 1998. 'The Status of Mechanism in Locke's *Essay*'. *Philosophical Review* 107, no. 3 (July): 381–414.

Lee, Sukjae. 2018. 'Causation'. In *The Routledge Companion to Seventeenth Century Philosophy*, ed. Dan Kaufman, 87–116. New York: Routledge.

Lee, Sukjae. 2019. 'Occasionalism'. In *Stanford Encyclopedia of Philosophy*, rev. 17 October 2019. https://plato.stanford.edu/entries/occasionalism/.

Locke, John. 1706. *An Examination of P. Malebranche*. In *Posthumous Works of Mr. John Locke*. London.

Locke, John. 1720. *Remarks on Norris*. In *A Collection of several Pieces of Mr. John Locke, Never before printed, or not extant in his Works.* ed. [Pierre Des Maizeaux]. London.

McCracken, Charles. 1983. *Malebranche and British Philosophy*. Oxford: Clarendon Press.

Malebranche, Nicolas. 1997. *Dialogues on Metaphysics and on Religion*, trans. Nicholas Jolley and David Scott. Cambridge: Cambridge University Press.

Nadler, Steven. 2011. *Occasionalism: Causation among the Cartesians*. Oxford: Oxford University Press.

Rozemond, Marleen, and Gideon Yaffe. 2004. 'Peach Trees, Gravity and God: Mechanism in Locke'. *British Journal for the History of Philosophy* 12, no. 3 (August): 387–412.

Sidelle, Alan. 2002. 'On the Metaphysical Contingency of Laws of Nature'. In *Conceivability and Possibility*, ed. Tamar Szabó Gendler and John Hawthorne, 311–36. Oxford: Oxford University Press.

Simpson, William, Robert Koons, and Nicholas Teh, eds. 2018. *Neo-Aristotelian Perspectives on Contemporary Science*. Routledge: New York.

Stuart, Matthew. 1998. 'Locke on Superaddition and Mechanism'. *British Journal for the History of Philosophy* 6, no. 3 (October): 351–79.

Wilson, Margaret. 1979. 'Superadded Properties: The Limits of Mechanism in Locke'. *American Philosophical Quarterly* 16 (April): 143–50.

Wilson, Margaret. 1982. 'Superadded Properties: A Reply to M.R. Ayers'. *Philosophical Review* 91 (April): 247–52.

Wood, Joshua. 2016. 'On Grounding Superadded Properties in Locke'. *British Journal for the History of Philosophy* 24, no. 5 (September): 878–96.

6

The Case against Powers

Walter Ott

It is, unfortunately, our destiny that, because of a certain aversion
toward light, people love to be returned to darkness.

Leibniz, 'Against Barbaric Physics'

Reading the contemporary literature on powers would be profoundly dispiriting
for the moderns. If there is one thing they agree on, it's the rejection of
Aristotelian powers.[1] In their eyes, to attribute such powers to bodies would
be to slide back into the scholastic slime from which they helped philosophy
crawl. We've had this argument before, they might say, and powers lost. If we
hope to resurrect something like the Aristotelian position, we had better be
sure we can answer the charges levelled by the moderns.

I don't propose to canvass every argument made in the modern period.[2]
Some of them, as we'll see, depend on views peculiar to the modern period
and so have lost their force. But one of them—Descartes's 'little souls'
argument—points to a genuine and, I think, persisting defect in powers
theories. The problem is that an Aristotelian power is intrinsic to whatever has
it. Once this move is accepted, it becomes very hard to see how humble matter
could have such a thing. It is as if each empowered object were possessed of a
little soul that directs it toward non-actual states of affairs and governs its
behaviour in actual ones.

---

[1] The antecedent might be false; Spinoza and even at times Leibniz himself might be exceptions to
the general hostility to Aristotelian powers. By 'Aristotelian' I mean any view that construes powers as
intrinsic properties of the objects that possess them.

[2] I set aside, for example, Malebranche's explicitly theological arguments (e.g. that attributing
powers to bodies is tantamount to paganism and should result in the worship of leeks and onions), as
well as arguments that depend on God's relation to the world (e.g. the divine concursus argument).
For Malebranche's complete slate of arguments, see Ott 2009. Contemporary objections to powers
theories that have no clear connection to the modern period also fall outside my purview. Perhaps the
best treatment of such objections is that of Psillos 2002.

Walter Ott, *The Case against Powers* In: *Reconsidering Causal Powers: Historical and Conceptual Perspectives*. Edited by:
Benjamin Hill, Henrik Lagerlund, and Stathis Psillos, Oxford University Press (2021). © Walter Ott.
DOI: 10.1093/oso/9780198869528.003.0007

When twentieth-century philosophers decided to resurrect the powers view, it was this original, Aristotelian theory they took as their inspiration.[3] I shall argue that this was a mistake. For in the early modern period we find not only profound arguments against Aristotelian powers but a way to preserve powers in the face of those same arguments. On the views of Robert Boyle and John Locke, powers are internal relations, not monadic properties intrinsic to their bearers. This move drains away the mysterious '*esse-ad*' or directedness of Aristotelian powers. And it solves what I'll suggest is the contemporary version of the little souls argument, Neil Williams's 'problem of fit'.

1. The Moderns' War on Powers

Although hostility to Aristotelian powers will come to be associated with British empiricism, particularly in the person of David Hume, the first clear statements of the case against powers come from René Descartes and Nicolas Malebranche. Some of these arguments seem very easy to bat away. Consider the charge that power attributions are explanatorily otiose: to say that opium puts people to sleep because of its *virtus dormitiva* is to contribute nothing by way of explanation. But, as others have pointed out, power attributions do exclude other explanations (it was the opium that did it, not watching TV). Nor are such attributions predictively vacuous: if the dormitive power brought about sleep, then taking opium again under similar circumstances will have the same result.[4]

These replies only take us so far, since they leave open that power attributions are temporary stopping places on the way to full explanation. What really undermines the vacuity charge is the progress of science, not conceptual analysis. For the early moderns, power attributions are obviously superfluous because there is a much better explanation on offer. When Descartes argues against powers in *Le Monde*, he does so only indirectly, by displaying the competing explanation in terms of the mechanical qualities: size, shape, and motion (AT 11:7–8; CSM 1:83). Power explanations might point you to the right actor on the scene (it was the opium and not the TV that did it), but they can hardly be the end of the matter when we have perfectly intelligible mechanical qualities to appeal to.

---

[3] As a case in point, consider Rom Harré and E. H. Madden's formulation: ' "X has the power to A" means "X (will) (can) do A, in the appropriate conditions, *in virtue of its intrinsic nature*" ' (1975, 86).
[4] Others who make these points include Nelson Goodman (1983), Stephen Mumford (1998, 136), and Brian Ellis (2002).

Unfortunately, the golden era of mechanism was short-lived.[5] Newton found himself obligated to postulate forces that could neither be explained by, nor reduced to, the mechanical properties of bodies. Although he is among the targets of Leibniz's 'Against Barbaric Physics' (Leibniz 1989, 312–20), Newton was no friend of scholastic powers and occult qualities (Henry 1994; Newton 2004, 103; Ott 2009, 7–9). Newton's solution is simply to bracket all question of the 'physical causes' and focus instead on the mathematics (Newton 2004, 63–4). The moderns' confidence that explanation in terms of size, shape, and motion trumps power explanations was misplaced.

In fact, it seems that many of the properties now postulated by fundamental physics are dispositional. It would be a mistake, I think, to conclude that there are no non-dispositional properties. The mere fact (if it is one) that physics finds no need to mention categorical properties is no evidence against their existence (Heil 2012, 64). And even if it were, pan-dispositionalism would still be a hasty conclusion to draw. For it follows, I think, only on assumption of a very strong reductivism. The fact that tables and chairs are made up of tiny bits doesn't on its own show that there are no tables and chairs. Similarly, even if the physics of very, very small things finds no need to invoke categorical properties, there might nevertheless be such properties at a higher level. Although science doesn't provide good reasons for going pan-dispositionalist, it does, I think, make it very hard to deny that there are dispositions. At a minimum, the moderns' quick declaration of victory was premature.

Alongside the arguments from explanatory and predictive impotence, we find a consistent line of thought that connects Descartes (AT 11:8; CSM 1:83), Malebranche, and Hume: the argument from nonsense. Malebranche puts it well: 'This opinion [that objects have powers] does not even seem conceivable to me. Whatever effort I make in order to understand it, I cannot find in me any idea representing to me what might be the force or the power that they attribute to creatures. And I do not even think it a temerarious judgment to assert that those who maintain that creatures have a force or power in themselves advance what they do not clearly conceive' (*Search*, Elucidation Fifteen, 658). The point cannot be that we are unable to form an image of a power, as we can of a key and a lock: that would be an odd argument for a rationalist like Malebranche to make. Instead, the point is that our conceptual resources

---

[5] In the *Principles*, for example, Descartes appeals to God's immutability to ground the laws of nature. Even Descartes, then, is not a *pure* mechanist, if that means that the only explanans in natural science are the size, shape, and motion of bodies.

do not suffice for the representation of powers. Although it's difficult to articulate, there does seem to be something fundamentally mysterious about powers. Once again, it is important to see Malebranche's argument in its intellectual context: when the standards of intelligibility are those of geometry, it is hard to deny that Malebranche has a point.

The moderns lodge a further charge against powers that has force even outside that context. This 'little souls' argument is best regarded as an extension of the argument from nonsense; unlike that argument, this one does not tacitly rely on a contrast with the perspicuous concepts of geometry.

In his Sixth Replies, Descartes describes his own youthful opinion of heaviness: 'I thought that heaviness bore bodies toward the center of the Earth as if it contained in itself some knowledge of it' (AT 7:441–2; CSM 2:297). This notion of heaviness as a kind of knowledge-directed striving, aiming at a goal outside of itself, is derived from the idea of the mind. It's minds, and minds alone, that act, strive, and know.

Picking up the same theme, Nicolas Malebranche offers this argument against the powers view: 'Well, then, let us suppose that this chair can move itself: which way will it go? With what velocity? At what time will it take it into its head to move? You would have to give the chair an intellect and a will capable of determining itself. You would have, in short, to make a man out of your armchair' (Malebranche 1992, dial. 7, 227). With an update in style, Malebranche might well have been the author of these words, written by Stephen Mumford in 1999: 'Does a soluble substance in any way strive to be dissolved? Does a fragile object aim to be broken? There is little reason to think that a material object without a mind is capable of having *aims* and *strivings* for events of a certain kind, because to do so would be for it to act, and attributions of action we reserve for things with minds' (Mumford 1999, 221; see also Armstrong 1999, 35). Mumford's remarks occur in the context of the debate over physical intentionality. Like minds, powers seem to have the ability to point to non-actual states of affairs: powers are directed toward their possible (even if never actual) manifestations. George Molnar (2003) and Ullin Place (1999) argue that such 'being toward' deserves to be counted as intentionality. Mumford, for his part, sees physical intentionality as bringing panpsychism in its wake: just like Descartes and Malebranche, Mumford thinks his opponents are making men out of armchairs.

In one way, the debate is something of a distraction from the real issue. The charge of panpsychism sticks only if intentionality is the mark of the mental. Since Place (Armstrong et al. 1996, 23) explicitly denies Brentano's thesis, he cannot be accused of imbuing the physical world with mental attributes (see also Molnar 2003, 70–1). As it turns out, the parties to this particular debate end up reconciling in an unexpected way: Place (1999, 231) demands to know just how Mumford's own theory avoids physical intentionality, and Mumford (2004, viii) 'leaves open' whether his functionalist theory of dispositions is 'really at odds' with Place's.

The real issue, I think, is not whether realism about powers requires animism. Instead, the question is how to understand the directedness that all powers seem to have. Even those who deny that this directedness is a form of intentionality seem to accept this feature of powers. Whether or not one wants to call it 'intentionality', part of the contemporary notion of a power, just like its scholastic antecedent, is its directedness to actual and possible manifestations.

The problem of directedness is a direct result of another feature of this concept of power. Nearly all parties accept that powers are intrinsic to the objects that have them. Roughly, the idea is that an object has the powers it does regardless of what else is going on around it. Fire has the power to burn paper, char flesh, evaporate fluid, change the colour of some kinds of materials, use oxygen, and on and on. Each power, all on its own, is poised to contribute its manifestation to all these events and more. Although part of the current orthodoxy, that feature of powers should strike us as prima facie very odd. It's as if fire knows just what to do whenever it comes across these things. And they, in turn, know just how to respond. Since the power is an intrinsic feature of the object that possesses it, it must come complete with a set of instructions, as it were, that tell it how to behave under every circumstance.

Prima facie oddity proves nothing, of course. I shall argue that the seventeenth-century 'little souls' objection becomes, in modern terms, what Neil Williams calls the 'problem of fit'. The real source of the problem is the Aristotelian's insistence on the intrinsicality of powers. We need, then, a tighter grip on just what that thesis amounts to.

2. Independence and Intrinsicality

Powers are independent of their manifestations. It is tempting to put the point by saying that a glass of water, for example, has the power to dissolve salt even if it never gets a chance to do so, indeed, even if there is no salt in the world to dissolve. To cash out independence in that way is to run it together with intrinsicality.[6] The two should be kept separate: the fact that a power can exist independently of its manifestations in no way shows that that power is intrinsic to the object that has it. I suspect that intrinsicality's claim to intuitive obviousness is entirely due to its tendency to masquerade as independence.

A commonly cited elucidation of the intrinsicality thesis comes from David Lewis: 'if two things (actual or merely possible) are exact intrinsic duplicates (and if they are subject to the same laws of nature) then they are disposed alike' (1997, 148). I think the proponent of intrinsicality should reject this formulation. By building in his parenthetical clause, Lewis has made powers extrinsic, by any reasonable definition. If the laws of nature vary, then salt that is water-soluble in the actual world is not so disposed in worlds that differ in the relevant laws. But that picture of laws is precisely what the powers view aims to undermine. For the powers theorist, laws are not 'top-down' features of the world that govern events. What 'governs' the course of nature is the powers. So, there is no sense to be made of intrinsic duplicates that are not disposed alike in virtue of being governed otherwise. The powers theorist, then, should prefer a simpler formulation: any intrinsic duplicates are dispositional duplicates.[7]

There is a final problem worth mentioning. As it stands, the intrinsicality thesis comes out as trivially true for pan-dispositionalism. If there is no such thing as a 'quiddity', or non-dispositional property, then to say that intrinsic duplicates are dispositional duplicates is just to say that any dispositional duplicates are dispositional duplicates. I am unsure how the pan-dispositionalist should refine the definition of intrinsicality.[8] In what follows, I will stick with the above formulation of the intrinsicality thesis, with the caveat that pan-dispositionalists will need to modify it.

The current orthodoxy, just like its scholastic counterpart, holds that all dispositions are intrinsic. The only major dissenter from that thesis is Jennifer McKitrick, and even she argues only that *some* dispositions are extrinsic. McKitrick (2003, 81) unearths only one argument for the intrinsicality of

---

[6] Molnar (2003) is one of the few writers who distinguish between the two.

[7] There is, of course, a debate over how to define 'intrinsic'; see, e.g., Langton and Lewis (1998). Molnar (2003, 39–43) has a very useful discussion of that issue. How that debate plays out will not, I think, affect the arguments of this chapter.

[8] An appealing, equally Lewis-inspired, definition is this: to say that object O has a power P intrinsically is to say that O would have P regardless of the other objects and powers in its world; indeed, O would have P even if there were no other objects in its universe. That sounds promising but notice that it's in danger of collapsing into the claim that O has P necessarily. And that would be a bad result, since an object's powers can come and go. Since I find pan-dispositionalism less than coherent, I am happy to plead the privilege of a sceptic.

dispositions, and it is less than impressive. Later, we will examine an argument of Molnar's that, in spite of his intentions, points us in the right direction.

3. The Problem of Fit

Neil Williams's (2010) 'problem of fit' threatens the whole project of reviving the moribund world of Aristotle. It is, if not just the little souls argument over again, clearly its contemporary descendant. The argument requires only a handful of assumptions. First, we have independence and intrinsicality. To this we add essentialism: the claim that the set of possible manifestations of a power is essential to it. In case it's not intuitively obvious, we can give a quick argument for it. Suppose there's some power P whose possible manifestations differ from one world to the next. It has one set of dispositions, call it Gamma, in this world, and another, Mu, in another world. But now it's open to us to take Gamma and Mu as subsets of P's total set of possible manifestations. So, there is no interesting sense in which the set of possible manifestations varies from one world to the next. If this argument goes through, I suspect it's because the essentialist thesis is an analytic consequence of the concept of a power.[9]

We need only one more ingredient: reciprocity. Powers come in pairs or bundles, not on their own. Most philosophers recognize the superficiality of distinctions like 'active' and 'passive'; the salt has to be able to cooperate in the dissolving every bit as much as the water has to be there to do the dissolving. The salt is not a mere background condition. As C. B. Martin puts it: 'I have been talking as if a disposition exists unmanifested until a set of background conditions is met, resulting in the manifestation. This picture is misleading, however, because so-called background conditions are every bit as operative as the identified dispositional entity. A more accurate view is one of a huge group of dispositional entities or properties which, when they come together, mutually manifests the property in question; talk of background conditions ceases' (Martin 2008, 50, quoted in Williams 2010, 87). Let's start with a simple example just to fix our ideas, abstracting from most members of this 'huge group': fire has the power to burn paper and paper has the power to be burned by fire. Since we have bought into intrinsicality, we cannot treat these claims

---

[9] I suppose someone with a governing concept of laws might deny this: the possible manifestations of a single power can vary with the laws of nature. But such a view is clearly antithetical to the powers theory, which is resolutely 'bottom-up'.

as merely two ways of saying the same thing. We have to be attributing two very different powers to the fire and the paper: the power to burn and the power to be burned.

But as Williams argues, we don't get this for free. Why should it be the case that paper's power is to be burned by fire, rather than to turn into a chicken when touched by fire, or to produce the sound of C#, or to pass along the flame to the next nearest object, or whatever you like? Powers come at least in pairs. Once we add in what we used to think of as background conditions, per Martin's instructions, we find that the situation becomes all the more puzzling. For it is not just *two* powers that have to co-occur; now there's an indefinitely large number of powers in play, all perfectly—and perfectly mysteriously—suited to their roles. Oxygen has the power to be consumed by flame and to enable it to do its work. It does not have the power, upon encountering the flame, to glow iridescent pink. And on and on.

Many, I assume, will react with incredulity: how can *this* possibly be a problem? Isn't it just obvious that powers 'cohere' with each other the way they do? I have some sympathy with this reaction. Indeed, my own preferred alternative will be to dissolve, rather than solve, the problem: at least one of the assumptions Williams made in generating the argument must be jettisoned. I do think that Williams is right in one respect: if you make those assumptions, you have to give a positive solution to the problem of fit. It is no good insisting that the powers obviously *do* fit. The question is, in virtue of what?

To that question, the power realist has no obvious answer. That powers 'fit' in the way they do looks very much like a miracle. A theist coming across the problem of fit would have new ammunition for a design argument: how else could it just happen to be the case that every power in our world is perfectly proportioned to all the others?

In fact, that way of putting the problem is a bit misleading. For it's at least plausible that the fit of powers is not a contingent matter. That seems to be Williams's (2010, 87; 89) position: 'the powers of the objects involved must cohere'; 'to fail to have the correct fit is to describe an impossible situation.' If anything, the necessity of fit makes the puzzle that much more pressing. The comparison with theism can be illuminating here. A design argument that appeals to the Earth's distance from the sun (a little bit further away, and it's too cold for life; a little bit closer, and it's too hot) is unconvincing for the obvious reason: the presence of life guarantees that the Earth is suitably situated to sustain life. But now suppose that the Earth's position were not contingent but necessary: in every possible world in which it exists at all, the Earth is just where it is. That really *would* call for a supernatural explanation.

So, we have two problems: why powers fit together as they do, and what makes it impossible for them to fail to do so. Although there are non-supernatural answers available, all of them either fail to solve these problems, or are enormously implausible, or both. The only way out, I shall argue, is to reject intrinsicality and so opt for the Boylean, not the Aristotelian, picture of powers.

4. Power Holism

The little souls objection points out the unappealing consequence of traditional powers ontologies: somehow, each power must have within it the means of pointing to all its possible manifestations. Whether powers so construed require us to be panpsychists is a side issue, or so I've argued. The real problem is that there's no way to stock each power with all the information it needs in order to behave appropriately.

The problem of fit is at bottom just the same problem, applied to pairs or n-tuples of powers rather than a power on its own. If we accept the presuppositions, we are, I think, stuck with a dramatically implausible conception of powers: what Williams calls 'power holism'. Here is Williams's statement of power holism: 'In order to provide the fit of powers, we must set up the powers so that they always match. How can this be done? One way is to cram all the information about every other property into the power, thereby "building" powers according to a plan—a plan that includes what kind of manifestation would result from each and every possible set of reciprocal partner.... each property contains within it a blueprint for the entire universe' (Williams 2010, 95). Given our assumptions, the only way to account for fit is to attribute to powers not just a directedness at some vaguely specified states of affairs but a full-blown and fully detailed recipe for every situation in which it might find itself. Fire doesn't *just* have the power to burn anything combustible; that is a superficial description of the power. Writing out the full specification of a power isn't even possible, since it has to include how it is to behave in the presence of every other possible power, whether that power is ever instantiated in its environment or not. Similarly, every other power has to include a description of how that power is to behave in the presence of the power(s) of fire.

Far from being unique to Williams, holism might be the default view among powers theorists. As Williams points out, Mumford seems committed to something like the view: for Mumford, 'the properties that are real in a world must...form an interconnected web' (2004, 182). In the work of other writers, it sometimes seems as if the problem of fit is being used to motivate holism, lurking just beneath the surface. When John Bigelow, Brian Ellis, and Caroline Lierse (2004, 158) argue that all laws of nature are due to the world's falling into the natural kind that it does, they do so partly because 'it is implausible that [the natures of the fundamental particles] should turn out to be independent of each other'. Surely one possible reason for thinking such independence implausible is the problem of fit.

We'll return in a moment to power holism. First, we should note that, even if it were defensible, it would not be a complete solution to the problem of fit. For the second question asks, in virtue of what do powers *necessarily* fit? It's not enough to say that each power is stocked with the appropriate information. There has to be some further reason why the powers that fit are the only ones that can get instantiated.

Williams opts for monism. Powers must fit together 'in virtue of their all being properties of the same particular' (2010, 100). The power to burn and the power to be burned are not properties had by discrete objects. If they were, it would be a mystery why they were instantiated in the same world. Instead, there is only a single object—the 'blobject'—that instantiates the whole set of properties, powers among them.

For my part, I cannot see how monism helps with this second question.[10] Why should being possessed by a single object explain the necessity of fit? Well, it might be the case that, if the powers did not fit, it would be impossible for them to be co-instantiated. That sounds right. But remember that the goal is to explain *why* it is impossible for non-cohering powers to occur in the same world. The explanation cannot merely be that they then would be conflicting powers had by the same object. For if it isn't impossible for such powers to be instantiated in the same world, then neither should it be impossible for them to be instantiated in the same object.

Here it is vital to keep in mind that, on the present view, the existence of conflicting powers is not prohibited by logical necessity. Given intrinsicality, there is no logical contradiction in a world that contains fire (with the power to burn paper) and paper (with the power to turn into a chicken on encountering fire).[11] That assumption is precisely what opens up the necessity gap: in

---

[10] I owe this point to Will Harris.

[11] It's not obvious how power holism purports to be consistent with intrinsicality. It seems that the holist has to deny that any monadic duplicate is going to be a dispositional duplicate. Here's one way around the problem: suppose that monism were indeed capable of securing the necessary agreement of the powers. Then there would be no possible world that duplicated an object from our world, for

virtue of what is such a state of affairs impossible? That gap cannot be closed by appealing to monism. More carefully: monism closes the gap only if one already assumes that conflicting powers cannot coexist in the same object. But there is no reason to think that, apart from the idea that such powers cannot populate the same possible world. And that idea is exactly what stood in need of justification. Monism does nothing to get us the necessity of the fit. The most pressing question, however, is the first: how is fit to be accomplished in the first place?

So, we should set aside the question of necessity and simply ask how power holism achieves fit and at what cost. As Mumford (2004, 215–16) and Williams (2010, 96) recognize, holism entails that if any one power were different, every other power would be different, too. This is a well-known problem for semantic holism, and it seems no more palatable in this context. Even prima facie unrelated powers are linked in a highly improbable fashion. Suppose Earth in world w1 contains water, with the power to dissolve salt. Earth in w2 is exactly like Earth in w1—it has salt, water, and so on—but w2 also has, in some unimaginably distant region of space, a single x-particle, with its own powers not instantiated anywhere in w1. Does w2 instantiate the power to dissolve salt? The holist has to say no. But this seems a very high price to pay. For salt, water, and everything else are just as they are on earth and throughout the Milky Way.

The powers theorist can of course say that water in w2 still has *some* power or other, among whose possible manifestations is the dissolving of salt. But whatever that power is, it is not the very same power that is instantiated in our world. There must be, in the space of possibilia, an infinite number of waters, each with its own distinct power-to-dissolve-salt, since all you have to do to generate new power is vary one other power, no matter how distant in time or space, no matter how causally isolated.

Williams isn't exaggerating when he says that every power has to contain 'a blueprint for the entire universe' (2010, 95). In this sense, powers are like Leibnizian monads: each one goes its own way under its own steam, and yet each one unfolds in accordance with all the others. Leibniz can tolerate this miracle because he has God in the picture to set up this pre-established harmony. Such an explanation for the fit of powers would be absurd, at least in the present context. And we have already seen that monism is no help in

example, and yet failed to duplicate all of the powers instantiated in our world, on pain of violating the necessity of reciprocity. So intrinsicality follows in the wake of the necessity of reciprocity, simply because you cannot duplicate one object with its powers without duplicating all of the powers in a world. Monism would then be the inevitable consequence of holism, not an optional add-on.

explaining this harmony: the same question arises whether the powers that be are instantiated in one object or many.

The comparison with semantic holism, which Williams himself draws, helps reveal other problems. It's natural to think that the meaning of a sentence is a function of the words that make it up (and what order they're in). The semantic holist cannot agree: on her view, no word has meaning in isolation. The same problem afflicts the powers holist: no power has a fixed nature apart from its co-instantiation with all other powers. How, then, are powers getting individuated? Power P's nature depends on those of powers Q and R, and so on, the buck always getting passed. That is at best unsatisfying and at worst question-begging.

But the chief weakness in power holism is, at bottom, just the little souls objection all over again. In its spirit, we should ask how exactly all the 'information' Williams speaks of is going to get 'crammed' into each power. 'Information' is ambiguous. Sometimes information means symbols with semantic content, in the sense that those symbols exist in a context of a convention that imbues them with meaning. In this sense, the phone book does, but grass does not, contain information. Presumably that cannot be what is meant. In other contexts, 'information' can mean anything that *could* be interpreted by someone as evidence of something. In this sense, grass contains information about recent weather patterns, the moisture in the ground, and so on. But this kind of 'information' is everywhere: every state of affairs can be grounds for *some* kind of inference. So that cannot be quite what is meant either.

Finally, one might think that the information required is something like a computer program, a set of if-thens that tell the power what to do in every possible state of affairs. (That powers contain information in this sense is, presumably, a consequence of Mumford's functionalist theory.) Programs in the literal sense require realizers: the 1's and 0's of your computer. Powers, however, are not themselves objects that admit of that sort of structure. So, once again, we have failed to cash out the relevant sense of 'information'.

One needn't be Malebranche, and still less Hume, to wonder whether power holism is intelligible. Another way to approach the same problem is in terms of a dilemma. Either a power is identical with its categorical base or it isn't.[12]

---

[12] In the latter class, I include views like those of Heil (2012) and Martin (2008) according to which categorical and dispositional qualities are different aspects of a single property. And, of course, pandispositionalists reject the demand for a categorical base at all and so fall, trivially, into the class of views that do not identify a power with its base.

Suppose that for a glass to be fragile is just for its particles to be arranged thus-and-so, and with such and such bonds among them. If so, then there is simply no way to cram all the necessary information into the available sack. Alternatively, suppose that the power is not identical with its base. So construed, being a power is a primitive notion in just the way that being a categorical property is. But now I have a still dimmer idea of what it could possibly mean to say that the power contains all the information it needs to behave itself appropriately.

5. Locks and Keys

If power holism is the cure, one might be forgiven for wanting the disease back. Although the assumptions that drive the problem of fit are commonly accepted today, one of them was much disputed in the seventeenth century: intrinsicality. For Boyle and Locke, among others, the scholastics' signal mistake was to treat powers as monadic properties of their bearers. I shall argue that this long-neglected path ought to be cleared of debris and followed.

Before beginning, it is worth thinking about what *kind* of solution the problem calls for. To settle that, one needs to decide what kind of necessity mandates the fitness of powers. Williams treats it as a kind of metaphysical necessity and then looks to monism to ground it. Something, I think, has gone wrong right at the start. It ought to come out as analytically true that any world contains all and only powers that fit.

It is only the artificial division among powers that generates the problem in the first place. We are invited to think of *the power of flame to burn paper* and *the power of paper to be burned by flame* as distinct properties, intrinsic to the things that have them. And of course, once one accepts this invitation, one has to explain why paper and fire instantiate just these and not any of the competing, incompatible powers we can think up. But surely at this stage one begins to suspect that there is something amiss with the problem. The distinction among these powers is an artificial one. Although it is perfectly fine to go on talking about *the* power to burn paper, what makes such statements true will not be an isolated property in the fire. Instead, it will be the whole complex of properties relevant to the event.

Making powers relational rather than intrinsic is a key step in the moderns' attempt to demystify them. In *The Origins of Forms and Qualities according to the Corpuscular Philosophy,* Robert Boyle[13] presents his case in terms of locks and keys:

> We may consider, then, that when Tubal Cain, or whoever else were the smith that invented *locks* and *keys,* had made his first lock...that was only a piece of iron contrived into such a shape; and when afterwards he made a key to that lock, that also in itself considered was nothing but a piece of iron of such a determinate figure. But in regard that these two pieces of iron might now be applied to one another after a certain manner...the lock and the key did each of them obtain a new capacity; and it became a main part of the notion and description of a *lock* that it was capable of being made to lock or unlock by that other piece of iron we call a *key,* and it was looked upon as a peculiar faculty and power in the key that it was fitted to open and shut the lock: *and yet by these attributes there was not added any real or physical entity* either to the lock or to the key, each of them remaining indeed nothing but the same piece of iron, just so shaped as it was before.
>
> (B 23, last emphasis mine)

If powers were intrinsic to their owners, then each time a lock was made that could be opened by a given key, the key would acquire a new, intrinsic property. But that is a mistake: what has happened to the key is a mere Cambridge change. The proposition 'the key has a new power' is surely true; but what makes it true is not a change in the key.

In fact, George Molnar (2003, 103–5) unwittingly directs us back to Boyle's view in his very argument for intrinsicality.[14] Responding to Boyle's argument, Molnar points out that what we might call the congruence of the lock and key is 'a comparative. Comparatives are founded relations that supervene on the properties of the relata...these properties have to include some that are part of the nature of the key and the lock respectively, and are therefore intrinsic to their bearers' (2003, 105). Molnar is, of course, right: the congruence of lock and key is a function of their intrinsic properties. Like internal resemblance—that is, resemblance in respect of intrinsic properties—congruence is a relation that one gets for free, as it were: fix the relevant properties of lock and key and congruence will ride in their train. That does nothing to show that

---

[13] For an illuminating discussion of these issues in Boyle's work, see Kaufman 2006. For a treatment of the ontology of relations in Locke and Boyle, see Ott 2017.

[14] At first Molnar seems to grant Boyle's point about locks and keys and insists that Boyle's claim cannot be generalized. But two pages later he attacks Boyle's example on its own merits. I focus on that argument, since I cannot make out why Molnar thinks the claim cannot be generalized.

the key's power to open the lock *is intrinsic to the key*, and that is what's in question. If it turns out that the power is founded on intrinsic properties of both the lock and the key, Boyle's withers will remain unwrung, for that is exactly what he is claiming.

6. The Positive View

The Boylean view claims that powers are not intrinsic to things that have them. 'The power to open locks of type-L' does not refer to a single property had by a given key, or even a given kind of key. It is at best a slightly misleading way of stating the facts of the case: this key, and others like it in relevant respects, are such that, when they are applied with sufficient force to this lock, and others like it in relevant respects, the lock is opened. It is no surprise that ordinary language has found ways to abbreviate this claim. But those linguistic shortcuts can't change the facts of the case.

The best way to develop this view is by considering objections. First, one might worry that it has Eleatic consequences. Relations, after all, depend on the existence of the relata. So, when all of the locks a given key fits are destroyed, the key loses its capacity to open these locks, even though it does not change in itself. The result is unappetizing: powers become 'mere Cambridge' properties, as Sydney Shoemaker (1980, 123) puts it.

But here it's vital to keep in mind that the power in question was never a power *of the key alone*. That's just what it means to deny intrinsicality. To be sure, if we insist on construing the power as belonging to the key alone, then the key gains and loses it depending on what locks there are. But that's not the view. The position claims that the power is a relation among all the relevant relata, however many there turn out to be. To make it the case that a power comes or goes, you have to alter or destroy at least one of the relata that make up the power relation. That's no more a mere Cambridge change than anything else. Consider an analogous worry: I can change the number of pages in a given book by ripping one or more of them out. The number of pages changes just in virtue of changing one of the pages, leaving the rest untouched. But the number of pages is not a mere Cambridge property for all that.

It remains the case, however, that an aggregate of relata can lose its congruence when one of those relata is destroyed. To take the sting out of this point, we have to deploy a distinction Shoemaker draws. He in effect distinguishes between powers as token-distinguished and as type-distinguished. What the key–lock aggregate loses when the lock is destroyed is the power P1 whose manifestation is the opening of lock 1 by key 1. That's the token version. By contrast, the type-identified power P2, defined as the power whose manifestation is the opening of locks of type-L by keys of type K, can still obtain in virtue of the relations in which those types stand. It's just this type/token confusion that allows Locke (*Essay* II.viii.19, 139) to say that porphyry in the dark has no colour, and Boyle to say that the key's capacity comes and goes.

Second, one might worry that the view invites a regress. If a power is a relation among, say, two objects or properties, in virtue of what does that relation obtain? Must there not be some further relation to guarantee the presence of this one? But as will already be clear, the regress has to assume that the relation in question is external. The Boylean view instead maintains that powers are internal relations: they obtain, if and when they do, in virtue of the intrinsic properties of the relata. What secures the presence of the relation is not some further relation but the properties of the things related.

This response invites a third objection: what exactly are the relata? They cannot themselves be powers, on pain of regress. But if they are not powers, they are, by definition, not capable of making a causal contribution to events. In short, we seem to have arrived at a view of powers that, paradoxically, deprives them of their 'oomph' (to use the technical term).

I think this objection begs the question against the relational view. Note that it has to assume without argument that anything that is not a power is causally irrelevant. There are general grounds to challenge this: as Brian Ellis (2010) has argued, spatio-temporal location clearly seems relevant to the exercise of powers but is not itself a power. But we don't need anything as sophisticated as that. On the present view, the power just is the relation: it is not reducible to any *one* of its relata. So, one cannot then complain that each relatum on its own is not itself powerful: that's just what the view claims.

But, the objector might insist, these relations are supposed to be internal. And internal relations are no addition to one's ontology. Compare internal resemblance: that a and b are similar in respect F just amounts to Fa and Fb. Whatever the relata turn out to be, the relation, and hence the power, will be nothing over and above them. To my ears, this objection sounds like a compliment, for it is this feature of the relational view that promises to remove the air of mystery.

Here a further clarification is in order. I have been talking as if the only intrinsic properties necessary for a power's exercise belong to the objects locally present when the event takes place. This is clearly false. At a minimum, some of those properties themselves will owe their existence to the exercise of other powers. (Gravity is a case in point.) As Locke puts it, 'Things, however absolute and entire they seem in themselves, are but Retainers to other parts of Nature' (*Essay* IV.vi.11, 587). Is the true subject of all power attributions, then, a single thing, namely, the world as a whole? (Bigelow et al. 2004).

To my ear, this question asks for a decision rather than a discovery. If the dependence of some causally relevant property P1 on the exercise of some further power P2 is enough to make one want to count P2 as one of the truthmakers for propositions about P1, that's perfectly fine. If one instead wants to count only the locally relevant properties (as defined by whatever scientific theory one is currently deploying), that's fine, too. The decision is a pragmatic one.

What more can be said about the intrinsic properties that together will make up the power? Not much, I think. That seems to be where philosophy ends and science begins. A host of other questions obtrude: are there powers at the macro level, the level at which Boyle and many other powers theorists work, that are not preserved at the micro level? If so, are those macro-level powers deprived of their ontological credentials? Are the relata that need to be included in any complete statement of a power so complex as to make a recurrence of that very same power extremely unlikely? These are questions I hope others persuaded of the merits of the relational view will take up.[15]

7. Conclusion

By way of summing up, we can ask, how does the Boylean view help with the problem of fit and the little souls problem generally? Recall that the problem of fit presupposes that we are dealing with powers intrinsic to a single actor. Why does the power of fire to burn paper never run across paper with the power to turn into a chicken when encountering fire? The Boylean view has a simple answer: there just is no such thing as *paper's* having the power to be burned or to turn into a chicken. Nor does a flame have the power to burn anything. What powers there are, are internal relations among intrinsic properties of things (or, for all I know, regions of space). We have given up on the whole project of somehow stocking each individual power with 'information' about its co-instantiates. And we have, as a result, got rid of the last remaining thorn of the little souls argument.

---

[15] It isn't clear that the Boylean view is as rare as I have made out. John Heil (2012, 148) takes causings, if not causal powers, to be internal relations. And Mumford himself (2004, 197) claims that 'causal powers' are internal relations. But as I read him, Mumford thinks that causal powers themselves are relations among further 'powerful properties'.

If the above arguments have any force, they suggest that the attempt to revive powers in contemporary philosophy made a mistake when it tried to go 'back to Aristotle'. To do so is to leap over the early moderns and their formidable case against powers. If there has indeed been some progress in philosophy since Aristotle, the natural point to which we should return is the early modern period, just before Hume's anti-realism about causation became all but inescapable.[16]

References

Armstrong, D. M. 1983. *What is a Law of Nature?* Cambridge: Cambridge University Press.

Armstrong, D. M. 1999. 'Comment on Ellis'. In *Causation and Laws of Nature*, ed. Howard Sankey, 35–38. Dordrecht: Reidel.

Armstrong, D. M., C. B. Martin, and U. T. Place. 1996. *Dispositions: A Debate*. Ed. Tim Crane. London: Routledge.

Bigelow, John, Brian Ellis, and Caroline Lierse. 2004. 'The World as One of a Kind'. In *Readings on Laws of Nature*, ed. John Carroll. Pittsburgh, PA: University of Pittsburgh Press, 141-60.

Ellis, Brian. 2002. *The Philosophy of Nature*. Montreal: McGill–Queen's University Press.

Ellis, Brian. 2010. 'Causal Powers and Categorical Properties'. In Marmodoro 2010, 133–42.

Goodman, Nelson. 1983. *Fact, Fiction, and Forecast*. 4th ed. Cambridge: Harvard University Press.

Harré, Rom, and E. H. Madden. 1975. *Causal Powers: A Theory of Natural Necessity*. Oxford: Basil Blackwell.

Heil, John. 2012. *The Universe as We Find It*. Oxford: Oxford University Press.

Henry, John. 1994. ' "Pray Do Not Ascribe That Notion to Me": God and Newton's Gravity'. In *The Books of Nature and Scripture*, ed. J. E. Force and Richard H. Popkin, 123–48. Dordrecht: Kluwer.

Kaufman, Dan. 2006. ' "Locks, Schlocks, and Poisoned Peas": Boyle on Actual and Dispositive Qualities'. In vol. 3 of *Oxford Studies in Early Modern Philosophy*. Ed. Daniel Garber and Steven Nadler, 153–98. Oxford: Clarendon Press.

Langton, Rae, and David Lewis. 1998. 'Defining "Intrinsic" '. *Philosophy and Phenomenological Research* 58, no. 2 (June): 333–45.

---

[16] I would like to thank Benjamin Hill, Henrik Largelund, and Stathis Psillos for their help.

Leibniz, Gottfried W. 1989. 'Against Barbaric Physics'. In *Philosophical Essays*, ed. and trans. Roger Ariew and Daniel Garber, 312–20. Indianapolis: Hackett Publishing Co.

Lewis, David. 1997. 'Finkish Dispositions'. *Philosophical Quarterly* 47, no. 187 (April): 143–58.

McKitrick, Jennifer. 2003. 'A Case for Extrinsic Dispositions'. *Australasian Journal of Philosophy* 81, no. 2: 155–74.

Malebranche, Nicolas. 1992. *Philosophical Selections*. Ed. Steven Nadler. Indianapolis: Hackett Publishing Co.

Marmodoro, Anna, ed. 2010. *The Metaphysics of Powers*. London: Routledge.

Martin, C. B. 2008. *The Mind in Nature*. Oxford: Oxford University Press.

Molnar, George. 2003. *Powers*. Oxford: Oxford University Press.

Mumford, Stephen. 1998. *Dispositions*. Oxford: Clarendon Press. Reissued with new preface, 2003.

Mumford, Stephen. 1999. 'Intentionality and the Physical: A New Theory of Disposition Ascription'. *Philosophical Quarterly* 49, no. 195 (April): 215–25.

Mumford, Stephen. 2004. *Laws in Nature*. London: Routledge.

Newton, Isaac. 2004. *Philosophical Writings*. Ed. Andrew Janiak. Cambridge: Cambridge University Press.

Ott, Walter. 2009. *Causation and Laws of Nature in Early Modern Philosophy*. Oxford: Oxford University Press.

Place, Ullin T. 1999. 'Intentionality and the Physical: A Reply to Mumford'. *The Philosophical Quarterly* 49, no. 195 (April): 225–31.

Psillos, Stathis. 2002. *Causation and Explanation*. Montreal: McGill–Queen's University Press.

Shoemaker, Sydney. 1980. 'Causality and Properties'. In *Time and Cause: Essays Presented to Richard Taylor*, ed. Peter van Inwagen, 109–35. Dordrecht: Reidel.

Williams, Neil E. 2010. 'Puzzling Powers: The Problem of Fit'. In Marmodoro 2010, 84–105.

7

The Return of Causal Powers?

Andreas Hüttemann

Powers, capacities, and dispositions (in what follows I will use these terms synonymously) have become prominent in recent debates in metaphysics, philosophy of science, and other areas of philosophy. I take their resurgence to consist in their acceptance as real and irreducible properties. This resurgence has been due to various kinds of developments; attempts to reduce talk about dispositions to talk about occurrences or categorical properties failed (see Mumford 1998; Schrenk 2016; McKitrick in this volume). While this in itself is not yet a positive argument for assuming dispositions to exist, various such arguments have been proposed (Shoemaker 1980; Mumford 1998; Ellis 2001; Molnar 2003; Bird 2007; for a classification see Bird 2016).

In this chapter I will comment on two rather different strands of argumentation for the acceptance of dispositions as real properties—one via Shoemaker and considerations of the identity of properties, another via Mill and Cartwright (the *extrapolation argument*). Shoemaker argued that 'the identity of a property is completely determined by its potential for contributing to the causal powers of the things that have it' (Shoemaker 1980, 133). This line of reasoning has been taken up by Ellis (2001) and Bird (2007), among others, identifying causal powers with dispositions (Chakravartty 2007, 123). In Section 1 I will briefly come back to this line of argument. In the bulk of the chapter (Section 2) I will examine a different line of argument for the existence of powers/capacities/dispositions. According to this argument the practice of extrapolating scientific knowledge from one kind of situation to a different kind of situation requires a specific interpretation of laws of nature, namely as attributing dispositions to systems. My main interest will be to discuss what characteristics these dispositions need to have in order to account for the scientific practice in question.

I will, furthermore, assess whether the introduction of dispositions in the context of the extrapolation argument can be described as a 'revitalization' or

Andreas Hüttemann, *The Return of Causal Powers?* In: *Reconsidering Causal Powers: Historical and Conceptual Perspectives*. Edited by: Benjamin Hill, Henrik Lagerlund, and Stathis Psillos, Oxford University Press (2021).
© Andreas Hüttemann.
DOI: 10.1093/oso/9780198869528.003.0008

as a 'return' to those notions repudiated by early modern philosophers (Section 3). More particularly, I will argue for the following four claims:

1. In repudiating scholastic terminology, including substantial forms with their causal powers, post-Cartesian philosophers focused on a concept of causation that was much stronger than twenty-first-century conceptions of causation. For this reason alone, whatever 'causal' is supposed to mean in today's causal powers, embracing causal powers is not a simple return to a pre-Cartesian notion.
2. The dispositions presupposed in scientific practice need not (and should not) be construed in causal terms (whether strong or weak).
3. While some early modern philosophers contrasted the characterization of the natural world in terms of substantial forms (and their causal powers) on the one hand and a mathematical characterization on the other, and suggested that these approaches are incompatible, the dispositions postulated by the extrapolation argument to account for scientific practice are themselves characterized in mathematical terms. More precisely: the behaviour the systems are disposed to display is—at least in physics—often characterized in mathematical terms.
4. The dispositions assumed in the law statements in scientific practice are determinable rather than determinate properties.

In what follows I will first have a brief look at the repudiation of substantial forms and their associated causal powers in early modern philosophy in order to argue for claim (1). This is relevant for assessing whether the causal powers argued for by Shoemaker and others should be considered as a return to those powers rejected by Cartesian philosophers (Section 1). Next, I will analyse the role laws of nature play in explanation, prediction, and other aspects of scientific practice. It will turn out that this role can best be understood by assuming that law statements attribute dispositional properties to objects or systems. In this context I will argue for claims (2)–(4) (Section 2).

1. The Rejection of Substantial Forms and Causal Powers

In the 13 July 1638 letter to Morin, Descartes argues that his assumption that matter is composed of extended parts explains a lot more than alternative conceptions of how bodies are constituted:

> You must remember that in the whole history of physics up to now people have only tried to imagine some causes to explain the phenomena of nature, with virtually no success. Compare my assumptions with the assumption of others. Compare all their real qualities, their substantial forms, their elements and countless other such things with my single assumption that all bodies are composed out of parts.... All that I add to this is that the parts of certain kinds of bodies are of one shape rather than another.... Compare the deductions I have made from my assumption—about vision, salt, winds, clouds, snow, thunder, the rainbow and so on—with what the others have derived from their assumptions on the same topics....I hope this will be enough to convince anyone unbiased that the effects which I explain have no other causes than the ones from which I have deduced them.
>
> (AT 2:99–100; CSMK 107)

Descartes contrasts two characterizations of nature: one in terms of scholastic terminology and substantial forms in particular, another in terms of matter conceived of as extended. The rejection of substantial forms came to be viewed as giving way to a mathematical characterization of the behaviour of bodies. Thus Newton—apparently appealing to a widely shared assessment—opens the 'Author's Preface to the Reader' of the *Principia* by describing the 'modern philosophers' as those philosophers who, after 'rejecting substantial forms and occult qualities, have undertaken to reduce the phenomena of nature to mathematical laws' (Newton 1999, 381).

The important point that I will come back to later is this: characterizing nature in terms of substantial forms (and their causal powers) was seen as a distinct and presumably incompatible approach to the modern approach of characterizing the phenomena in terms of mathematical laws.

So, the repudiation of (among other things) substantial forms and their causal powers allowed for a mathematical physics to be developed. Another consequence of the repudiation of substantial forms, however, pertains to the notion of causation. The active powers of substances were supposed to be grounded in and unified by their substantial form (see Des Chene 1996, 157–67 for discussion of some views on the exact relation between substantial form and active powers). The rejection of substantial forms and active (or causal) powers thus meant that the very concepts in terms of which causation was explicated were no longer available. While Descartes nowhere explicitly discusses this implication, post-Cartesian philosophers acknowledged it—either by downplaying its significance (La Forge, Cordemoy) or embracing it somewhat enthusiastically for theological reasons (Malebranche).

This discussion of the implication of the repudiation of substantial forms is interesting because it indicates that the concept of causation that was at stake cannot easily be equated with anything we discuss today. One initial piece of evidence is provided by the well-known fact that Malebranche assumes that between a cause and an effect a necessary connection has to obtain. This assumption was not made up by Malebranche to attack a straw man (as is sometimes assumed) but rather refers back to a notion of causation as explicated for example by Suárez (see Ott 2009 for discussion). Suárez was interested in whether 'there are causes that act necessarily once the things required for acting are present' and explains:

'This question is easy and so one should assert succinctly, that among created substances there are many that operate necessarily once all the things they require for operating are present. This is obvious from experience and from a simple induction. For the sun illuminates necessarily, and fire produces warmth necessarily, and so on for the others. The reason for this must stem from the intrinsic and determination of [the agent's] nature'

(DM 19.1.1, 1:688a; Suárez 1994, 270).

And further:

Lastly one can infer from this that the necessity in question is so strong that neither the intrinsic power of the faculty itself nor any other natural cause whatsoever is able to remove it or to prevent it from issuing in an act. To be sure, natural causes can, as we have explained, impede one another through resistance or through contrary action, and in this way they are capable of removing all the things that are required for acting. But once those things are posited, natural causes cannot prevent the action of a necessary agent, since they do not have the power either to change the nature of things or to remove wholly intrinsic properties....even God himself does not seem to be able to bring it about...that a cause which by its nature acts necessarily should fail to act, once all the things required for acting are posited.

(DM 19.1.14, 1:692b; Suárez 1994, 281)

Given that Malebranche, in disputing the causal efficacy of created substances in his *The Search after Truth*, refers to Suárez quite frequently (for example, at least five times in Elucidation Fifteen on the efficacy of secondary causes), it is interesting to note that Suárez relies on a concept of causation that is presumably influenced by Neoplatonic thought. Suárez defines the concept of causation (which is meant to comprise all four Aristotelian causes) as follows: 'a cause is a principle per se inflowing being to something else [*Causa est Principium* per se *influens esse in aliud*]' (*DM* 12.2.4, 1:384b) This characterization had the explicit purpose not only of unifying the various Aristotelian notions of causation but also of bringing causation and creation under a common heading (see Schnepf 2006, 230–57). With this background and the explicit characterization (to inflow being [*esse*] into something else) in mind, it is maybe less surprising than it might initially appear that Malebranche concludes that, if we allow the causal efficacy of secondary substances, 'we therefore admit something divine in all the bodies around us when we posit forms, faculties, qualities, virtues, or real beings capable producing certain effects through the force of their nature' (Malebranche 1997, 446). Given a concept of causation as *inflowing being* [*esse*] *into something else*, to cause something may indeed appear to be a divine activity.

Further evidence for the claim that we cannot simply equate the notion of causation (and a fortiori that of a causal power) that came under attack in early modern philosophy with contemporary notions can be gained from a passage in Louis de La Forge. La Forge, while not referring back to Suárez explicitly, argues, just as Malebranche, that strictly speaking only God is able to cause effects in other substances: 'However you should not say that it is God who does everything and that the body and mind do not really act on each other. For if the body had not had such a movement, the mind would never have had such a thought, and if the mind had not had such a thought the body might also never have had such a movement' (La Forge 1997, 150). Thus, for La Forge, the rejection of causation among created substances in a strict or strong sense is compatible with a weaker notion of cause. According to Specht, La Forge allows for 'quasi-causation' (Specht 1966, 140). This weak notion of causation (quasi-causation) can quite naturally be construed as a counterfactual conception of causation. So, what was at stake when the traditional causal terminology was rejected by early modern philosophers wasn't one of our contemporary conceptions of a causation.[1] In rejecting substantial forms and their causal powers, early modern philosophers attacked a notion of causation that was much stronger than any of ours, namely that of

---

[1] I think it can easily be argued that a regularity account of causation is compatible with the absence of strong causation too. Even the transfer of conserved quantities as conceived in transfer theories of causation can be explicated in terms of the continuous recreation of amounts of energy at different places in space and is thus compatible with occasionalist views. La Forge and Malebranche had different aims in defending occasionalist positions. While La Forge tended to downplay counterintuitive implications of Cartesianism, Malebranche welcomed, for example, the passivity of created substances. While La Forge endorsed quasi-causation, Malebranche presumably would not have liked to do the same. However, it seems that causation according to a counterfactual account, according to a regularity account, and even according to a transfer theory account is compatible with his position.

causation as the inflow of being into something else. As a consequence, even if the strand of argumentation according to which 'the identity of a property is completely determined by its potential for contributing to the causal powers of the things that have it' (Shoemaker 1980, 133) convincingly establishes the reality of causal powers, it is doubtful whether the 'revitalization' of causal powers should be viewed as a simple return to the very conception that was rejected by Cartesian and post-Cartesian philosophers because these latter causal powers were conceived of in terms of 'inflowing being to something else'.

2. The Role of Laws and Dispositions in Scientific Practice

I will now examine what I take to be the most convincing argument for dispositions: an argument that appeals to the role of laws of nature in scientific practice, namely in extrapolation. In Section 3 I will consider—with respect to dispositions as introduced by this argument—the question whether the introduction can be described as a 'revitalization' or as a 'return' to those notions repudiated by early modern philosophers.

2.1. Law Statements and Generalizations

The argument from extrapolation to the existence of dispositions analyses the role law statements play in this context. It is thus important to begin with a clear understanding of what laws of nature (law statements) are.

Let me start with Galileo's law. It may be thought that Galileo's law is simply identical to the following equation: (1) $s = \frac{1}{2} gt^2$ (where s is distance covered, t is time, and g is a constant). That seems wrong to me. It is fairly uncontroversial to take laws or law statements to be those (maybe complex) generalizations that play a role in extrapolation, confirmation, explanation, and other aspects of scientific practice. With this characterization of a law statement as a starting point, we can immediately infer the following consequence: if a law statement is what is confirmed or disconfirmed in trials (or used in the contexts of explanation, prediction, or manipulation), an equation on its own cannot be an example of a law (or a law statement—in what follows I will use these two terms synonymously). As a matter of fact, nobody takes Galileo's law to be disconfirmed by balls uniformly rolling on a horizontal plane or by stones lying on the ground, both of which fail to satisfy equation (1). What is missing is a claim about *the kinds of systems* that are meant to be represented by the equation. Galileo's law is not simply a mathematical equation. Nor does it suffice to add that t represents time and s the path taken by an arbitrary object. Galileo's law is the claim that the behaviour of a particular class of systems can be represented by the above equation. A full statement of Galileo's law might thus be something like the following: (1') *Free falling bodies* behave according to the equation $s = \frac{1}{2} gt^2$.

Similarly, $F = ma$ is merely a mathematical equation. It becomes a law statement once it is asserted that this equation is meant to represent the behaviour of physical systems, indeed, of all physical systems whatsoever. And again, the Schrödinger equation with the Coulomb potential on its own does not qualify as a law statement; that is, it is not what we confirm or disconfirm. By contrast, the claim '*Hydrogen atoms* behave according to the Schrödinger equation with the Coulomb potential' is a law statement.

The fact that equations such as $s = \frac{1}{2} gt^2$ come with a domain of systems for which they are meant to be relevant has been noted by others, such as within the semantic account of theories. Thus, Bas van Fraassen, referring to Ronald Giere, defines a *theory* (not a law[2]) as consisting of (a) the *theoretical definition*, which defines a certain class of systems, and (b) a *theoretical hypothesis*, which asserts that certain (sorts of) real systems are among (or related in some way to) members of that class (van Fraassen 1989, 222). A preliminary general characterization of law statements might thus be the following:

(A) All systems of a certain kind K behave according to Σ.

Here 'Σ'—the law predicate—typically stands for an equation or a set of equations. The expression 'of a certain kind K' may refer to all physical systems whatsoever, as it does in the case of Newton's second law or in the case of the bare Schrödinger equation. Or it might refer to a more circumscribed class of systems such as free-falling bodies or hydrogen atoms, thus giving rise to so-called *system laws*.[3] It is important to note that the behaviour attributed to the systems in question is in general complex and relational. In the case of free-falling bodies, the length of the path and the time taken are related, not only for actual values of the variables but for all possible (or some restricted domain of) values. Taking 'All ravens are black' as a paradigm for law statements ignores the complex structure usually attributed to systems.

---

[2] In fact, Giere (1995, 120–38) and Van Fraassen (1989, 183–214) deny that there are laws of nature. According to my reconstruction, what Giere and Van Fraassen call a 'theory' should be taken to be a law statement.

[3] One might worry about the exact characterization of the system to which Σ is attributed. The worry is that one needs Σ to individuate the systems in question. That, of course, would make the law statement an analytical truth and thus devoid of empirical content. It has to be assumed that the relevant class has been individuated antecedently, for example in terms of experimental procedures ('free-falling bodies'), or by other means that do not depend on Σ. This is a thorny issue that I will not go into in this chapter.

Another example that illustrates the structure of law statements is Euclidean geometry. Euclidean geometry on its own is a mathematical theory without any empirical import. We get an empirically testable claim (a law) if we take a certain class of systems (space-times) to be adequately characterized in terms of Euclidean geometry.[4]

Characterizing law statements in terms of (A) allows me to draw attention to an important distinction between different kinds of generalizations. Even though there are often no explicit quantifiers, law statements often involve at least *two different kinds* of generalizations. In Galileo's law, for example, we can distinguish one form of generalization that pertains to the values of the variables *s* and *t* that appear in the equation and another generalization that pertains to the objects to which a certain kind of behaviour is attributed. More generally we can distinguish the following two kinds of generalizations (see Scheibe 1991 for this distinction)[5]:

(a) *System-internal generalizations*—generalizations concerning the values of variables, for instance *s, t* in the case of Galileo's law; the equation may hold either for all values of the variables or for all values within a certain range;

(b) *system-external generalizations*—generalizations concerning different systems; the equation pertains to all systems of a certain kind (for example, free-falling bodies).

The essential point is that law statements attribute predicates to systems (via external generalizations), and these predicates typically involve a highly complex mathematical apparatus, which implies internal generalizations. This complexity becomes invisible if we are operating with examples of the 'All ravens are black' sort.

With this preliminary characterization of law statements, I can now turn to the argument for dispositions.

2.2. Dispositions

Braveness, fragility, and solubility are usually considered to be dispositional properties. By contrast, squareness and other geometrical properties are

---

[4] This point was famously observed by Einstein (1983, 28–9): 'As far as the laws of mathematics refer to reality, they are not certain; and as far as they are certain, they do not refer to reality.' Einstein's own view is fairly close to what has been suggested here.

[5] Hitchcock and Woodward (2003, 189) draw attention to this distinction—though not in these terms—when they remark with respect to explanation that 'the nomothetic approach has focused on a particular kind of generality: generality with respect to objects or systems other than the one whose properties are being explained'. By contrast, their own account of explanations relies on generalizations that pertain to the values of variables.

often considered to be the paradigmatic candidates for categorical properties. It has turned out to be notoriously difficult to draw a distinction between dispositional and categorical properties (see Mumford 1998; Choi and Fara 2018). In this chapter I will try to get along with assumptions about dispositional and categorical properties that are fairly uncontroversial.

A useful starting point is Mathew Tugby's characterization of dispositionalism and dispositions:

> According to dispositionalism, dispositions (or what are sometimes called 'causal powers') are taken to be real properties of concrete things, and properties which cannot be reduced to any more basic kind of entity. But what, precisely, does it mean to say that a property is irreducibly dispositional in nature? Roughly, it means that the property is characterized in terms of the causal behaviour which things instantiating that property are apt to display. This is to say, in other words, that irreducibly dispositional properties are by their very nature orientated towards certain causal manifestations. To illustrate: in order to explain what it means for, say, a particle to have the property of being charged, the dispositionalists will point out for example that charged particles accelerate when placed in an electro-static field. In explaining this feature of charge, dispositionalists take themselves to have said something about the essential nature of charge. In short, then, dispositionalists see properties as properties for something else: their causal manifestations.   (Tugby 2013, 452)

This is a useful starting point because it allows me to start with a disagreement: there is no need to consider the relation between a disposition and its manifestation as causal. While 'causal' appears in Tugby's *abstract* characterization of dispositions ('irreducibly dispositional properties are by their very nature orientated towards certain causal manifestations'), the example for a disposition (charge), by contrast, is characterized without relying on causal terminology: an object has the property of being charged, if—when placed in an electrostatic field—it accelerates in the right way. It seems to me that the three occurrences of the word 'causal' in the above quote can be skipped without loss of content. I will, thus, work with a notion of a disposition that does not explicitly build causation into it. Assumptions about causation are additional assumptions and there is no need to build them into a characterization of dispositions. (There is a further reason for this approach: as I argued elsewhere (Hüttemann 2013), causation can be explicated in terms of dispositions. For this to be non-circular, dispositions should better be explicable without recourse to causal terminology.)

As a consequence, I start with a slightly modified version of Tugby's characterization: a dispositional property is characterized in terms of the behaviour which things instantiating that property are apt to display. Dispositional properties thus allow for a distinction to be drawn between a property being instantiated and a property being manifest, while non-dispositional or categorical properties do not allow for this distinction. Dispositional properties allow for this distinction because they are 'oriented towards their manifestation', as Tugby notes, which implies that the manifestation need not be displayed.[6] Thus a person can be brave without actually displaying brave behaviour. By contrast, a piece of paper cannot instantiate the property of squareness without displaying squareness. It seems inappropriate to consider such a distinction in the case of squareness. If we furthermore assume that the distinction between categorical and dispositional properties is exhaustive, we can derive a simple and fairly uncontroversial sufficient condition for the ascription of dispositional properties: if the ascription of a property assumes that the property in question can be instantiated without the manifestation being displayed, it is a dispositional property. Whatever the exact nature of dispositions may be, this criterion will allow us to identify some dispositions. It is all we need for the conclusions of this chapter.

2.3. The Problem of Extrapolation

In their 2012 book Nancy Cartwright and Jeremy Hardie (2012, 4) discuss the following question: a project carried out in Tennessee in the 1980s showed, according to an empirical evaluation, that students in smaller classes did better than students in larger classes and that this applies in particular to minority students. In the mid-1990s California intended to apply a class-size reduction program to its state schools. Will what has worked in Tennessee work in California too? This question concerns the issue whether evidence gained in one case is relevant for a distinct case in another context.

---

[6] While I think this way of talking is fairly common, some authors (Molnar 2003, 195; Mumford 2009, 104) argue that dispositions *contribute to effects*. According to these authors, dispositions *always* contribute and, furthermore, the contributions remain the same (though the effects may differ, provided different dispositions contribute to an effect). These contributions are then identified as the dispositions' manifestation, and as a consequence these dispositions are always manifest (for discussion see McKitrick 2010). However, this does not undermine the above criterion, for Molnar and Mumford would agree that a disposition ascription allows for a disposition to be instantiated but without *the effect* to which the dispostion tends occurring. Thus, with respect to our criterion these differences are largely terminological.

The problem Cartwright and Hardie are concerned with is the question of finding evidence for whether or not we can infer from one case to the other. However, there is a further issue: assume that the inference works; in virtue of what does it work? The problem of extrapolation, as discussed in this chapter, consists in the challenge to explain why generalizations that we have found to hold under specific circumstances hold under different circumstances too. (See Steel 2007, 3, for examples and a slightly different characterization of the problem.)

The problem of extrapolation is relevant not only for social policy or in medical research but also in very simple examples from physics, to which I will now turn. Take Galileo's law again. It describes the behaviour of free-falling bodies; it describes the behaviour of a body falling in a vacuum. What about falling bodies in a medium? As a matter of fact, we consider the vacuum case to be relevant for the other cases too: we take the vacuum case as the basis of our theoretical treatment of the other cases and add further influences (further terms). So, as a matter of fact we are assuming that what we know about one kind of, in some way ideal, case is relevant for other, less ideal, cases, for other cases with different contexts. Why can we extrapolate from one case to the other?

Cartwright in another publication suggested the following explanation of what is going on in extrapolation: 'When . . . disturbances are absent the factor manifests its power explicitly in its behaviour. When nothing else is going on, you can see what tendencies a factor has by looking at what it does. This tells you something about what will happen in very different, mixed circumstances—but only if you assume that the factor has a fixed capacity that it carries with it from situation to situation' (Cartwright 1989, 191). In my terminology (Hüttemann 2014), Cartwright argues that we need dispositions to understand why extrapolation works. With respect to Galileo's law the problem for a non-dispositionalist reading of laws can be formulated as follows: in the case we hold that Galileo's law applies to the vacuum case only (that is how we have reconstructed the law up to this point by restricting it to *free*-falling bodies), we cannot explain why it is also used for accounting for the behaviour of falling bodies outside the vacuum/the laboratory. If we assume that Galileo's law applies to falling bodies in general (we drop the 'free' in 'free-falling bodies'), we might seem to be in a better situation, because it is now clear why the law is relevant both in and outside the laboratory/vacuum. However, the problem is that for falling bodies outside the vacuum it is false that they fall according to the equation $s = \frac{1}{2} gt^2$. On a non-dispositionalist reading, Galileo's law is either irrelevant for falling bodies outside the vacuum or false.

Cartwright's way out is the way out that has already been suggested by Mill:

There are not a *law* and an *exception* to that law—the law acting in ninety-nine cases, and the exception in one. There are two laws, each possibly acting in the whole hundred cases, and bringing about a common effect by their *conjunct operation* [my emphasis].... Thus if it were stated to be a law of nature, that all heavy bodies fall to the ground, it would probably be said that the resistance of the atmosphere, which prevents a balloon from falling, constitutes the balloon as an exception to that pretended law of nature. But the real law is that all heavy bodies *tend* to fall.   (Mill 2008, 56)

To return to the case of extrapolation: when—in a successful case of extrapolation—we consider Galileo's law ('disturbances are absent...nothing else is going on') to be relevant for a falling body in a medium ('very different, mixed circumstances'), we assume that something carries over from the first (ideal) situation to the second situation—a property that is (completely) manifest in the first situation but fails to be (completely) manifest in the second (though instantiated in both situations).

If we consider the fall in the vacuum to be, for example, explanatorily relevant for the fall in the medium, we rely on how the body would behave if it were isolated, that is on how the falling stone would behave in the absence of the medium. But the body, in this case, is not isolated. The behaviour to which the body is disposed is not (completely) manifest due to the presence of the medium—however, the body and the medium *contribute*[7] to the behaviour of the compound, and the respective dispositions are thus *partially* manifest (I will say a few more words about partial manifestation below).

So the claim is that, if we take the external generalizations to be attributing dispositions (capacities), we understand why the law statement can be extrapolated, for example, to systems in non-ideal situations, that is, in the presence of further factors. Something (a dispositional property) is present in both situations. The fact that we are dealing with a property (a universal) accounts for the fact that what we know in the one situation is relevant to the other situation too. The fact that the property is dispositional has to be

---

[7] This notion of contribution needs to be distinguished from the notion introduced by Molnar (2003) or Mumford (2009). According to the terminology used here, a disposition can be either completely manifest (if the context is ideal) or partially manifest. In the latter case it is said to contribute to the behaviour that is due to various factors or dispositions. So, contributions are not conceived of as additional ontological entities mediating between disposition and effect/manifestation.

assumed because in (at least) one of these cases the property isn't (completely) manifest.[8]

But how *exactly* does the ascription of dispositions help to explain why we may extrapolate, for example, from an ideal case to a less ideal case? The worry is that citing a disposition might not explain the *actual* behaviour of a system. Earman and Roberts, for example, inquire: 'Thus if what one wants explained is the actual pattern, how does citing a tendency—which for all we know may or may not be dominant and, thus, by itself may or may not produce something like the actually observed pattern—serve to explain this pattern?' (Earman and Roberts 1999, 451–2). The fact that something (the disposition) *carries over* does not tell us enough as long as it isn't clear *how* the disposition contributes to the phenomenon in question/the behaviour that arises in new contexts in which (further) interfering factors are present.

However, if we know how to handle interfering factors, the claim that dispositions contribute to the 'actual pattern' or the observed behaviour can be made precise. Consider again the simple case of a falling body. Within Newtonian mechanics we can describe what is going on in a system consisting of the earth's gravitational field and the body within this field. It might be argued that knowledge about how this system (and the body as part of it) would behave if it were on its own does not tell us anything about how it actually behaves in the presence of other factors, for example a medium. For all we know, it might be argued, the other factors might or might not be dominant. But things are not as gloomy as this objection implies. We know how to determine what is happening in such cases. We have the means to determine *how (to what extent) the various factors add up or contribute*. When we are considering a falling body in a medium, we have to add, for example, an extra force term into Newton's second law, which represents buoyancy. The essential point is that there are laws or rules that allow us to quantitatively determine how different factors or tendencies contribute to the actual behaviour that we want to explain.

These laws of composition also allow us to make the notion of partial manifestation precise: a disposition is *partially* (as opposed to *completely*) manifest if it contributes to some behaviour given the presence of other contributing

---

[8] It should be noted that, while the usual examples of dispositions such as fragility and solubility need positive triggering conditions to become manifest, in our case the manifestation conditions for complete manifestation are negative. It is required that the system in question (free-falling body) is isolated or on its own. (This would be different if we were considering examples from biology or social policy. In these cases we would have to hold a lot of features constant rather than absent.)

dispositions. Laws of composition tell us what happens if various dispositions are present, and thus 'partial contribution' can be explicated and made quantitatively precise.

Earman and Roberts's worry is unwarranted as long as there are laws of composition that allow us to estimate the contribution of various factors. In physics there are various quantitatively precise generalizations for this purpose, such as those describing the vector addition of forces, or more generally rules that describe how to represent compound systems in terms of subsystems by way of laws (rules) of composition (Hüttemann 2004; 2015).

Let me summarize. When we successfully extrapolate from one situation to a qualitatively different situation, we assume something, a property of the system in question, to be present in both situations that accounts for the extrapolation. However, this property needs to be dispositional because if the situations are qualitatively different in the sense that the observed behaviour that is to be explained is different, then in at least one of the cases the property is not (completely) manifest. A fortiori the properties that account for extrapolation are dispositional properties according to the criterion discussed in Section 2.2.

This result has consequences for how law statements should be understood. Those law statements that describe the behaviour of systems which may or may not be completely manifest should be taken to attribute dispositions to systems. In the case of falling bodies we should reconstruct Galileo's law as follows: falling bodies *are disposed to behave* according to the equation $s = \frac{1}{2} gt^2$. More generally, to account for the role law statements play in scientific practice, we should reconstruct them as making the following claim: All physical systems of a certain kind *are disposed to behave* according to Σ.

3. Revitalization

So, the argument from extrapolation shows that dispositions should be accepted as real. Is this a return to or a revitalization of concepts that have been repudiated by (some) early modern philosophers? Given the foregoing consideration, we certainly see a return of dispositions, that is of properties that allow for a distinction between being instantiated and being manifest. However, for the following reasons it is not a return to a conception of causal powers that was renounced, for example by occasionalists.

First, while I have given no positive characterization of the relation of a disposition to its manifestation, the argument from scientific practice I looked at provides no reason to take this relation to be a *causal* relation. There is no need to conceive the dispositions that account for extrapolation in causal terms. It is a return of powers, not of causal powers.

Second, what the disposition is disposed to, its manifestation, is at least in physics typically characterized in terms of mathematics; in simple cases such as Galileo's law in terms of a single equation, and in more complex cases in terms of Maxwell's equations or the Einstein equations. While some early modern authors suggested that a characterization of nature in terms of substantial forms and causal powers is incompatible with a characterization in terms of mathematics, the dispositions introduced in the context of the extrapolation argument are typically disposed towards the display of a behaviour that is characterized in terms of mathematics.

Third, let me add a further and final point that shows why we should hesitate to describe the return of dispositions as a simple revitalization of older notions. Law statements, as we have seen, attribute dispositions to systems. In contrast to, say, the power to generate fire or the power to illuminate, these dispositions are typically (infinitely) multitrack. In virtue of the internal generalizations, a law statement (like ideal gases behave according to the equation $pV = vRT$) implies an infinite number of regularity statements for every single system. For example: for every ideal gas, if $p = p_0$ and $V = V_0$, there will be a value T_0 for T such that $p_0 V_0 = vRT_0$; for every ideal gas, if $p = p_1$ and $V = V_1$, there will be a value T_1 for T such that $p_1 V_1 = vRT_1$; for every ideal gas, if $p = p_2$ and $V = V_2$, there will be a value T_2 for T such that $p_2 V_2 = vRT_2$; and so on. The fact that laws such as the ideal-gas law are formulated in terms of functional dependencies (and the internal generalizations that come along with them) shows that the dispositions in question are explicitly multitrack. Even though there are some discussions about whether a *multitrack* or *determinable* disposition can be a fundamental feature of reality (Bird 2007; Armstrong 2010; Wilson 2012), it is a characteristic of those dispositions attributed to systems by law statements. While the characterization of a *power to illuminate* may be vague and thus implicitly allow for different ways of illuminating, the characterization of a disposition in terms of variables and functional dependencies is *explicitly* multitrack. Thus, the fact that law statements attribute *determinable* or *infinitely multitrack* dispositions to systems is a further reason to be wary to take contemporary reference to dispositions to be a return to those notions repudiated in early modern philosophy.

4. Conclusion

In the previous sections I examined two different strands of argumentation for the reality of powers, capacities, or dispositions, one via Shoemaker and considerations of the identity of properties, another via the extrapolation argument. With respect to both lines of argument I argued that the powers, capacities, or dispositions these arguments seek to establish do not constitute a return to those causal powers that have been rejected by some early modern thinkers.

One reason is that the notion of causation that is constitutive for the causal powers rejected by some early moderns is much stronger than anything that is envisaged in today's causal powers by property theories that follow Shoemaker's argument. The notion of disposition or power relevant for the extrapolation argument doesn't even require the powers to be causal powers at all. Furthermore, these dispositions are multitrack and their manifestation needs to be characterized in terms of mathematics (two features not pertinent to traditional causal powers).

References

Armstrong, D. M. 2010. *Sketch for a Systematic Metaphysics*. Oxford: Oxford University Press.

Bird, Alexander. 2007. *Nature's Metaphysics*. Oxford: Oxford University Press.

Bird, Alexander. 2016. 'Overpowering: How the Powers Ontology Has Overreached Itself'. *Mind* 125, no. 498 (April): 341–83.

Cartwright, Nancy. 1989. *Nature's capacities and Their Measurement*. Cambridge: Cambridge University Press.

Cartwright, Nancy, and Jeremy Hardie. 2012. *Evidence-Based Policy: A Practical Guide to Doing It Better*. Oxford: Oxford University Press.

Chakravartty, Anjan. 2007. *A Metaphysics for Scientific Realism: Knowing the Unobservable*. Cambridge: Cambridge University Press.

Choi, Sungho, and Michael Fara. 2018. 'Dispositions'. In *Stanford Encyclopedia of Philosophy*, revised 22 June 2018. https://plato.stanford.edu/entries/dispositions/.

Des Chene, Dennis. 1996. *Physiologia: Natural Philosophy in Late Aristotelian and Cartesian Thought*. Ithaca: Cornell University Press.

Earman, John, and John Roberts. 1999. 'Ceteris Paribus, There is No Problem of Provisos'. *Synthese* 118, no. 3 (March): 439–78.

Einstein, Albert. 1983. 'Geometry and Experience'. In *Sidelights on Relativity*, 27–56. New York: Dover.

Ellis, Brian. 2001. *Scientific Essentialism*. Cambridge: Cambridge University Press.

Giere, Ronald. 1995. 'The Skeptical Perspective: Science without Laws of Nature'. In *Laws of Nature*, ed. Friedel Weinert, 120–38. Berlin: de Gruyter.

Hitchcock, Christopher, and James Woodward. 2003. 'Explanatory Generalizations, Part II: Plumbing Explanatory Depth'. *Noûs* 37, no. 2 (June): 181–99.

Hüttemann, Andreas. 2004. *What's wrong with Microphysicalism?* London: Routledge.

Hüttemann, Andreas. 2013. 'A Disposition-based Process Theory of Causation'. In *Metaphysics and Science*, ed. Stephen Mumford and Matt Tugby, 101–22. Oxford: Oxford University Press.

Hüttemann, Andreas. 2014. 'Ceteris Paribus Laws in Physics'. In 'Ceteris Paribus Laws Revisited', ed. Alexander Reutlinger and Matthias Unterhuber. Supplement, *Erkenntnis* 79, no. 10 (December): 1715–28.

Hüttemann, Andreas. 2015. 'Physicalism and the Part-Whole-Relation'. In *Metaphysics in Contemporary Physics*, ed. Tomasz Bigaj and Christian Wüthrich, 51–72. Amsterdam: Rodopi.

La Forge, Louis de. 1997. *A Treatise of the Human Mind*. transl. by D. Clarke. Dordrecht: Kluwer.

Malebranche, Nicolas. 1997. *The Search After Truth*. transl. by Thomas Lennon and Paul Olscamp, Cambridge: Cambridge University Press.

McKitrick, Jennifer. 2010. 'Manifestations as Effects'. In *The Metaphysics of Powers: Their Grounding and Their Manifestations*, ed. Anna Marmodoro, 73–83. New York: Routledge.

McKitrick, Jennifer. 2021. 'Resurgent Powers and the Failure of Conceptual Analysis'. in this volume.

Mill, J. S. 2008. 'On the Definition and Method of Political Economy'. In *The Philosophy of Economics: An Anthology*, ed. Daniel Hausman, 3rd ed., 41–58. Cambridge: Cambridge University Press.

Molnar, George. 2003. *Powers*. Oxford: Oxford University Press.

Mumford, Stephen. 1998. *Dispositions*. Oxford: Clarendon Press.

Mumford, Stephen. 2009. 'Passing Powers Around'. *Monist* 92, no. 1 (January): 94–111.

Newton, Isaac. 1999. *The Principia: Mathematical Principles of Natural Philosophy*. Trans. I. Bernard Cohen and Anne Whitman, assisted by Julia Budenz. Berkeley and Los Angeles: University of California Press.

Ott, Walter. 2009. *Causation and Laws of Nature in Early Modern Philosophy*. Oxford: Oxford University Press.

Scheibe, Erhard. 1991. 'Predication and Physical Law'. In 'Predication: Analyses and Implications', ed. Karel Lambert and Alan Code. *Topoi* 10, no. 1 (March): 3–12.

Schnepf, Robert. 2006. *Die Frage nach der Ursache: Systematische und problemgeschichtliche Untersuchungen zum Kausalitats- und zum Schopfungsbegriff.* Göttingen: Vandenhoeck & Ruprecht.

Schrenk, Markus. 2016. *Metaphysics of Science: A Systematic and Historical Introduction.* New York: Routledge.

Shoemaker, Sydney. 1980. 'Causality and Properties'. In *Time and Cause: Essays Presented to Richard Taylor*, ed. Peter van Inwagen, 109–35. Dordrecht: Reidel.

Specht, Rainer. 1966 *Commercium mentis et corporis: Über Kausalvorstellungen im Cartesianismus.* Stuttgart-Bad Cannstatt: Friedrich Frommann Verlag.

Steel, Daniel. 2007. *Across the Boundaries.* Oxford: Oxford University Press.

Suárez, Francisco. 1994. *On Efficient Causality: Metaphysical Disputations 17, 18, and 19.* Trans. Alfred J. Freddoso. New Haven: Yale University Press.

Tugby, Matthew. 2013. 'Platonic Dispositionalism'. *Mind* 122, no. 486 (April): 451–80.

Van Fraassen, Bas. 1989. *Laws and Symmetry.* Oxford: Clarendon Press.

Wilson, Jessica. 2012. 'Fundamental Determinables'. *Philosopher's Imprint* 12, no. 4 (February): 1–17.

8

Qualities, Powers, and Bare
Powers in Locke

Lisa Downing

Locke's version of the distinction between primary and secondary qualities has long been taken as especially distinguished, despite his clear debts to Boyle, Descartes, Galileo, and others. Nevertheless, debate still continues on the precise form that the distinction actually took for Locke. Indeed, the last decade has seen intense interest in Locke's metaphysics, including prominent new analyses of the metaphysics of the distinction in Locke by scholars including Jacovides (2017), Ott (2009), Pasnau (2011), and Stuart (2013). Here I will critically consider some of this recent work on the way to defending answers to three questions: (1) What is it to be a primary quality, for Locke? (2) How are primary and secondary qualities related exactly? And, most controversially, (3) what status do secondary qualities have for Locke? What can we say about their ontology? With regard to (3), I question some recent interpretations that see Locke as advocating a kind of anti-realism about secondary qualities, as well as interpretations that label him as some sort of reductionist about secondary qualities. I argue that we need to consider Locke's epistemology of qualities in order to draw the right conclusions about his ontology of qualities.

1. What Kind of Distinction is the Primary/Secondary Quality Distinction?

Let's begin, however, with a prior, more basic question: what kind of distinction is the primary/secondary quality distinction? Mi-Kyoung Lee helpfully identifies two different ways of understanding the distinction itself:

Lisa Downing, *Qualities, Powers, and Bare Powers in Locke* In: *Reconsidering Causal Powers: Historical and Conceptual Perspectives*. Edited by: Benjamin Hill, Henrik Lagerlund, and Stathis Psillos, Oxford University Press (2021).
© Lisa Downing.
DOI: 10.1093/oso/9780198869528.003.0009

Thus, the primary–secondary quality distinction can be understood in two ways: (1) it is a way of marking off a metaphysical distinction between essential and non-essential properties of matter and of bodies. As such, it promises to be a basic feature of any materialist ontology, and hence one would expect any theory of matter to have commitments on such a question. (2) In another sense, it is a way of marking off those sensible qualities which seem to be particularly subjective, that is, dependent on the responses that perceivers have to them. Qualities like colors and flavors give rise to conflicting appearances in different perceivers, and this in turn seems to have something to do with the epistemic facts about our sense-modalities and modes of perception.   (Lee 2011, 17)

This looks like two different ways of generating the distinction and identifying which qualities are in which category. On the one hand, you might think the essential or foundational qualities are the key to the distinction and are identified first, via one's views about matter. On the other hand, you might think that there is a direct route to the secondary qualities by locating the qualities that are subjective, perceiver-dependent. Thus, somewhat crudely, version (1) of the distinction is matter-driven, while version (2) of the distinction is subjectivity-driven.

In this chapter, I will consider the distinction only as it is found in the early modern period, and my focus will be on Locke. I follow the mainstream in contemporary Locke scholarship in supposing that (1) is the best way to understand Locke's distinction. I think this is clearly also true of Galileo, Descartes, and Boyle: they have views about what body is fundamentally like; those views are views about which qualities are primary; and that's what is driving their versions of the distinction. They then need an account of where the other apparent qualities come from. That account makes crucial reference to appearance and the senses. So, on this version (1) story, secondary qualities are leftovers: they are the directly observable qualities that aren't primary. The story told about these qualities may make them subjective in some way, but it's primary qualities that come first in the account.

Nowadays, this is the mainstream view both of Locke's version of the distinction as well as of the distinction in the early modern period generally, and that's for an obvious reason: it has become commonplace to view the distinction as having emerged from new mechanist conceptions of matter put forward in the scientific revolution by Galileo, Descartes, Boyle, etc. This mainstream view is clearly correct, yet it should be noted that seeing the distinction, even in the early modern period, as subjectivity-driven is not without significant textual support, for where the distinction is supposed to be motivated by arguments from illusion or relativity, there the distinction seems to be subjectivity-driven.

Thomas Hobbes, for example, uses relativity arguments (as pointed out by Pasnau 2011, 510–12): 'And to proceed to the *rest* of the *senses*, it is apparent enough, that the *smell* and *taste* of the *same thing*, are *not* the *same* to *every man*; and therefore are not in the thing *smelt* or *tasted*, but in the men' (Hobbes 1962, 1:8). Pierre Bayle, as is well known, portrays the distinction as driven by relativity arguments, though he also prefigures Berkeley in noting the weakness of such arguments: 'Add to this, that all the ways of suspension which destroy the reality of corporeal qualities, overthrow the reality of extension. Since the same bodies are sweet to some men, and bitter to others, it may reasonably be inferred that they are neither sweet nor bitter in their own nature, and absolutely speaking.... Why should we not say the same thing of extension?' (Bayle 1734–8, 5:612). And Locke, notoriously, tells us about the water cold to one hand, warm to another:

> *Ideas* being thus distinguished and understood, we may be able to give an Account, how the same Water, at the same time, may produce the *Idea* of Cold by one Hand, and of Heat by the other: Whereas it is impossible, that the same Water, if those *Ideas* were really in it, should at the same time be both Hot and Cold. For if we imagine *Warmth*, as it is *in our Hands*, to be *nothing but a certain sort and degree of Motion in the minute Particles of our Nerves, or animal Spirits*, we may understand, how it is possible, that the same Water may at the same time produce the Sensation of Heat in one Hand, and Cold in the other.... But if the Sensation of Heat and Cold, be nothing but the increase or diminution of the motion of the minute Parts of our Bodies, caused by the Corpuscles of any other Body, it is easie to be understood, That if that motion be greater in one Hand, than in the other; if a Body be applied to the two Hands, which has in its minute Particles a greater motion, than in those of one of the Hands, and a less, than in those of the other, it will increase the motion of the one Hand, and lessen it in the other, and so cause the different Sensations of Heat and Cold, that depend thereon. (*Essay* II.viii.21, 139)

However, the consensus view of Locke's distinction as matter-driven and influenced by mechanism has a convincing account of what is going on in this last, famous passage. We should appreciate that Locke is exhibiting the *explanatory power* of the mechanist version of the doctrine of primary and secondary qualities. With the corpuscularian hypothesis, we may *give an account* of the phenomena, which phenomena include conflicting appearances. This interpretation fits very well with the texts, while making clear that, on this interpretation, the primary–secondary quality distinction is version (1), matter-driven.[1] While it is considerably less clear what exactly to say about Hobbes, Bayle, and others, more generally we might say that, regarding observations about relativity, conflicting appearances may be brought in in order to exhibit what a matter-driven version of the distinction can explain, or even as a sort of independent confirmation of the (version 1) doctrine.

In what follows, I will take for granted that the distinction, for Locke, is matter-driven, as it quite clearly is for many in the seventeenth century. Thus, primary qualities come first. But what exactly is a primary quality, and how should we identify these qualities?

2. How Should we Define and Delimit Primary Qualities?

We can start here with the suggestion from Lee: recall that Lee characterizes version (1) of the distinction as holding that primary qualities are the qualities that are *essential* to matter or body (Lee 2011, 17). Lee's characterization seems remarkably well suited to Galileo, especially in his early (1623) articulation of the distinction in *The Assayer*. Recall a central passage of this famous discussion:

> as soon as I conceive of a corporeal substance or material, I feel indeed drawn by the necessity of also conceiving that it is bounded and has this or that shape; that it is large or small in relation to other things; that it is in this or that location and exists at this or that time; that it moves or stands still; that it touches or does not touch another body; and that it is one, a few, or many. Nor can I, by any stretch of the imagination, separate it from these conditions. However, my mind does not feel forced to regard it as necessarily accompanied by such conditions as the following: that it is white or red,

---

[1] Rickless (2014) is an illustrative recent exception to the near-consensus interpretive view of Locke's distinction as a version (1) distinction. For Rickless, secondary qualities really do come first for Locke, and primary qualities are defined relative to them. He holds that for Locke secondary qualities are perceiver-dependent qualities, and the relativity arguments are arguments that a set of qualities are perceiver-dependent. So, they are at the foundation of the distinction. He holds that primary qualities are 'real' qualities for Locke, but that 'real' is a technical term meaning 'not perceiver-dependent' (89). Rickless acknowledges that this saddles Locke with relying on a number of unpersuasive arguments (92).

bitter or sweet, noisy or quiet, and pleasantly or unpleasantly smelling; on the contrary, if we did not have the assistance of our senses, perhaps the intellect and the imagination by themselves would never conceive of them. Thus, from the point of view of the subject in which they seem to inhere, these tastes, odors, colors, etc., are nothing but empty names; rather they inhere only in the sensitive body, such that if one removes the animal, then all these qualities are taken away and annihilated.   (Galilei 2008, 185)

Galileo singles out the primary qualities as those our minds feel forced to regard as in the corporeal body. That the other qualities are not conceptually necessary to body he reinforces with the suggestion that we wouldn't conceive of them if we lacked sensation. If we suppose that Galileo is making the assumption, common in the period, that conceptual necessity tracks necessity, then it looks like primary qualities are just essential qualities, those that bodies cannot lack and still be bodies.[2]

Let's next turn to Locke, to see how the suggestion that primary qualities are the essential qualities of bodies might be applied. For Locke, this suggestion immediately raises the question: are we talking about nominal essences or real essences? If the answer is 'real essences', we might wonder why we can be so confident in supposing that we can identify the primary or essential qualities, given Locke's epistemic modesty and inclination towards extreme pessimism about our knowledge of particular real essences.

Matthew Stuart (2013, 46–52) gives the alternative answer—'nominal essences'. His conceptualist interpretation[3] of Locke posits that the primary qualities of bodies are just those that we take to be required in order to be body, that is those that we've enshrined in the nominal essence of body. This view is well motivated in that it nicely explains why Locke is so confident in his particular list of primary qualities—size, shape, solidity, and motion-or-rest. If we are simply identifying the content of our idea of body, the genus body as we have defined it, then we can easily make a list of primary qualities via the sort of reflection reported by Locke in the famous grain of wheat passage (*Essay* II.viii.9, 135).

I submit, however, that the conceptualist account of primary qualities faces serious problems. The most straightforward is that when Locke discusses primary qualities, it seems pretty clear that he is discussing the nature of reality

---

[2] There is room to doubt, however, that this is in the end the best characterization of what it is to be a primary quality for Galileo, as I indicate in note 7 below.

[3] Stuart borrows the contrast between conceptualist and transdictive interpretations from Robert Wilson (2002), who defends a transdictive view.

rather than our classificatory schemes.[4] This is most obvious where he declares that primary qualities '*are really in them,* whether any ones Senses perceive them or no' (II.viii.17,137) as opposed to those which are '*imputed*' (II.viii.22, 140) and 'nothing in the Objects themselves, but Powers' (II.viii.10, 135). Further evidence that Locke connects his notion of primary qualities to reality, rather than to our conceptual schemes, can be found in the way he connects them to 'modifications of matter' (II.viii.7, 134) and describes himself as having detoured into 'Physical Enquiries' (II.viii.22, 140).[5] Thus, it seems that primary qualities should be logically connected to the real essence of matter, not merely to its nominal essence. Turns out, there is evidence that Locke sees this connection; in IV.vi.7 he writes, 'we know not the real Constitutions of Substances, on which each *secondary Quality* particularly depends' (582). 'Real constitution' is systematically used by Locke as synonymous with 'real essence'. He says further in IV.vi.7 that secondary qualities depend on real essences. But, of course, Locke usually describes secondary qualities as depending on primary qualities. This highlights the logical relationship between real essence and primary quality.

A further problem faced by Stuart is that on his interpretation, as on most interpretations, *secondary quality* is a metaphysical notion. (As we will see shortly, it is for Stuart (2013, 117–21) the metaphysical notion of a degenerate power.) But, for Locke, all the qualities that are directly perceivable, that is all the qualities that correspond directly to our simple ideas, are either primary or secondary. The apparent qualities of bodies divide into primary and secondary. Thus, on Stuart's interpretation, in forming our concept of body we would have to have successfully identified all the qualities that aren't actually degenerate powers. But (1) this seems implausibly optimistic, and (2) it entails that primary qualities aren't merely conceptually primary, since they are the apparent qualities that aren't degenerate powers. That is to say, an interpretation of primary qualities as *merely* conceptual is not consistent with a metaphysical interpretation of secondary qualities.[6]

---

[4] Admittedly, our classificatory schemes derive from experience, and experience derives from reality, so the conceptualist needn't deny all connection to reality. Still, I argue below that Locke is supposing a tighter connection than that.

[5] Thanks to Henrik Lagerlund for this point.

[6] Possibly Stuart would be content to merge the conceptual and the transdictive views, since he is mostly concerned to give a conceptualist interpretation of II.viii.9. However, (1) this would undercut one motivation for the conceptual interpretation, and (2) the result would be that primary qualities cannot simply be defined in terms of the contents of our idea of body, because they have to be more than just that.

For a recent interpretation of Lockean primary qualities which grounds them in metaphysics, we can turn to Michael Jacovides (2017), who identifies primary qualities as the qualities that are *propria* of matter. He sees Locke as following in a scholastic tradition here: 'Locke defines primary qualities as those that "are utterly inseparable from the Body, in what estate soever it be" (2.8.9). The definition makes them the *propria* of body, where *proprium* is taken in something like Sanderson's second sense: a property of A is a feature that all As have at all times' (88). This (rightly, in my view) connects primary qualities to the way bodies actually are. Nevertheless, I suggest that it isn't optimal as a definition, for Locke.

Suppose that some bodies have a type of basic quality that cannot be causally derived from some underlying quality or qualities but that this quality isn't found in all body or matter. I submit that Locke ought to and would regard such a quality as primary. He doesn't consider the possibility because he generally assumes—following Boyle, who builds it into the content of the corpuscularian hypothesis—that matter is 'catholick' (Boyle 1999–2000, 5:305), that is, that it is everywhere the same. He ought, however, to regard that as an empirical issue, and his commitment to catholick matter is less deep than his commitment to a metaphysics of primary qualities. Thus, we ought not to define primary quality in a way that builds in this assumption.[7] Jacovides's *propria* interpretation falls short by taking inseparability the wrong way. Primary qualities are inseparable or ineliminable for Locke not because they must belong to all matter as such, but because they aren't derived and so can't be eliminated, where they occur, by some sort of decomposition.[8]

The most Lockean of recent treatments of Locke's notion of primary quality, in my view, is Robert Pasnau's (2011). Pasnau interprets Locke as intending, first and foremost, 'a thesis about what is explanatorily basic in the natural

---

[7] For what it's worth, I think this could well be true of Galileo also. Suppose there were some quality that didn't occur in all bodies, and thus wasn't essential to body, but was both foundational and efficacious where it did occur. I suggest (speculatively, I admit) that Galileo would count it as a primary quality. That the primary qualities are essential qualities is (or ought to be) a hypothesis for Galileo: checking for conceptual necessity is a way of identifying the minimal set of primary qualities, and Galileo proposes (with, admittedly, more confidence than seems entirely justifiable) that that minimal set will suffice.

[8] I am arguing, then, that being a *proprium* in Jacovides's sense, aka Sanderson's second sense, isn't a necessary condition for being a primary quality, for Locke. Jacovides (2017, 73) also connects *propria* to following from the essence, again in keeping with the traditional scholastic view. But this makes clear that the *propria* interpretation connects Locke too firmly to a metaphysics that he has largely left behind: the essence of body and matter *as a kind*, for Locke, is a nominal essence—the abstract idea of a thing with a particular set of observable qualities. The real essence of matter is then whatever constitution produces that set of qualities. That real essence could be disjunctive, and all sorts of qualities could follow from it.

world' (489; Jacovides 2017, 86). Like Jacovides, Pasnau elaborates by connecting Locke to scholasticism: Pasnau holds that the intension of the notion of primary quality for the scholastics is captured by the tripartite formula: universality, supervenience, and causal primacy. He holds that the moderns follow this intension (while replacing the scholastic *extension* of hot, cold, wet, and dry), concluding that 'the essence of Locke's primary–secondary quality distinction is an empirical claim consisting in the theses of explanatory priority, universality, and causal primacy' (Pasnau 2011, 488).

He labels this as an empirical claim, and this requires clarification. I take it that he means that it is an empirical claim that the mechanist list of qualities—size, shape, motion or rest, solidity—satisfy the criteria of universality, causal primacy, and explanatory priority. I agree with this, though, as I have argued elsewhere (Downing 1998; 2013), I think that it is a hypothesis that eventually Locke sees reason to distance himself from. That is to say, Locke has a metaphysical notion of primary quality, and it is in the end an empirical question which qualities are in fact primary (though Locke thinks we have some reason to privilege mechanism as a hypothesis about the answer to this empirical question).

Pasnau notes that universality floats free of his other two criteria, and this seems clearly right. As we just saw in considering the *propria* interpretation, it can only be an empirical claim that the qualities that have causal primacy and explanatory priority are also universal. The mechanists are all inclined to assume this, but they ought to be prepared to revise that assumption if necessary. It is not, however, an ordinary empirical claim for Locke that some qualities have causal primacy and explanatory priority. Rather, it is part of the basic metaphysical picture that he expects particular physical theories to fill in.

Once universality is omitted, I'm in close agreement with Pasnau's analysis, though it is unclear what reason there is to separate explanatory priority and causally primacy: if qualities have causal primacy, they presumably must be used in full explanations, and if the only relevant explanation here is causal explanation, then explanatory priority brings with it causal primacy. I suggest the formula 'intrinsic and foundational' to characterize primary qualities. If a quality must, in principle, be used in explanation, then it is foundational. And if it is foundational, it cannot be explained away, and so must be used (at least) in explaining itself. I include 'intrinsic' as well in order to try to capture an assumption that I take to be typical of the moderns and to be deeply held by them—that extrinsic properties are always explicable in terms of intrinsic properties plus something like spatial position. These intrinsic and foundational qualities are the qualities that characterize the real constitutions of particular bodies, because they ground their further powers (as in *Essay* IV.vi.7, 582).

3. From Primary Qualities to Secondary Qualities

Another question still to be addressed is, if secondary qualities are to be defined in relation to primary qualities, then how so? The rough idea here is that secondary qualities are leftover qualities: the apparent qualities that aren't primary must be secondary. But this requires some refining.

One notorious textual problem in Locke that motivates some of the refining is posed by a passage where Locke seems to tell us that *all* qualities are powers:

> Whatsoever the Mind perceives in it self, or is the immediate object of Perception, Thought, or Understanding, that I call *Idea*; and the Power to produce any *Idea* in our mind, I call *Quality* of the Subject wherein that power is. Thus a Snow-ball having the power to produce in us the *Ideas* of *White*, *Cold*, and *Round*, the Powers to produce those Ideas in us, as they are in the Snow-ball, I call *Qualities*; and as they are Sensations, or Perceptions, in our Understandings, I call them *Ideas*.   (*Essay* II.viii.8, 134)

This passage is much debated; some commentators discount it, some build interpretations around it (Stuart 2013, 36; Jacovides 2017, 186). I think it should not be discounted, but it can be readily enough explained. Locke is considering here the qualities that correspond to our ideas, that is to say, *observable* or *macroscopic* qualities. Those qualities are all powers—powers to produce ideas in us. Locke's commitment to this claim is evident also in his treatment of the adequacy of our simple ideas, a key epistemological thesis to which we will have reason to return.

All observable, macroscopic qualities of bodies, the qualities that correspond one-to-one with ideas, are powers. The question of which of the observable qualities (if any) are primary is for Locke the question of which of those qualities are *more than mere powers*, that is, which correspond to ideas that provide us with an accurate conception of the way bodies are in themselves. Locke appeals here to his notion of resemblance. Some of our ideas of bodies may 'resemble' or accurately represent the foundational, intrinsic properties of bodies. More specifically, some of our ideas may give us an accurate conception of the types of properties that are intrinsic and foundational in bodies.

Other ideas, by contrast, threaten to mislead us. The qualities corresponding to the 'resembling' ideas count as primary qualities, those which do not are secondary. Thus, both the intrinsic, foundational properties of bodies, which might belong only to submicroscopic parts and so be unobservable, and those macroscopic qualities, which are powers, and thus causally dependent on more fundamental qualities, that provide us with an accurate conception of the intrinsic, irreducible properties, count as primary qualities (for discussion see Downing 1998, 390).

The category of primary quality is thus at the level of quality-instances a disjunctive one, because macroscopic primary qualities are powers to produce ideas in us, while submicroscopic ones are not. The sorts of qualities that are intrinsic and foundational in bodies are the primary qualities, those that aren't are secondary. What secondary qualities are are powers, *mere* powers because they don't resemble (give us an accurate conception of) the qualities that are actually intrinsic and foundational in bodies.

Secondary qualities, then, are mere *powers*. How should we understand this claim? This is a question that I will not attempt to reach the bottom of. I argue for two main points in what follows: contra several recent interpreters, (1) we *should* understand Locke as putting forward a kind of *realism* about secondary qualities, and (2) we *should not* be quick to assume that Locke is endorsing *reductionism* about secondary qualities.

4. The Ontology of Secondary Qualities

4.1. Realism

First, let's establish a bit of context. There are many different answers given or suggested in the seventeenth century to the question, what are secondary qualities such as colours? Possible answers would seem to include: colours are ideas or sensations, states of sense organs, arrangements of particles on the surfaces of objects, relations between objects and perceivers, or powers in objects to produce ideas in us. Indeed, one can find bits of text in Boyle (1999–2000, 5:309–10; B 23–4) that suggest almost all of these positions, though his focus is on the last two, as is brought out by his famous extended comparison of sensible qualities with a key's ability to open a lock. Descartes's official position (AT 8A:322–3: CSM 1:285; AT 7:82: CSM 2:56–7; Schmaltz 1996, 56; Menn 1995, 206 n35) seems to be that terms like 'red' are equivocal: we can use them to talk about sensations in us, which are modes of mind, or we can use them to talk about an unknown configuration of the parts of bodies that is the source of that sensation.

Locke is most commonly applauded for putting secondary qualities unambiguously in bodies: ideas are in us but qualities are in bodies—they are powers that bodies have to produce ideas in us. This has often been characterized as a form of realism about secondary qualities, where the contrast is presumably with idealism about secondary qualities. Nevertheless, a number of recent interpretations of Locke either explicitly or implicitly attribute anti-realism about secondary qualities to him, according to which secondary qualities either do not exist or are radically mind-dependent. In this section, I consider several such recent interpretations, and the relations among them, with the aim of exhibiting their considerable disadvantages, as either views about secondary qualities such as colour or as interpretation of Locke.

Jacovides (2017, 192–5) forthrightly embraces an anti-realist interpretation: secondary qualities are not real beings, aren't entities of any kind in bodies, and thus do not exist. Instead, on his view, Locke provides us with a *semantics* for secondary quality terms, which allows us to keep talking about colours, despite their irreality. This view gets some textual motivation from Locke's emphatic declarations that it's the primary qualities that are *really* in bodies. To counter Jacovides, however, when Locke tells us that secondary qualities are *nothing* in objects, he typically elaborates, explaining that they are nothing in objects *but powers*: mere powers, or bare powers, that is, powers whose resulting ideas mislead us about what the objects are actually like. It is true that Locke would deny that these powers are *res*, items separable from what they qualify in the manner of scholastic real qualities. And it sometimes sounds as if this plausible point is the one Jacovides wants to make in stating that Locke denies that secondary qualities are 'real beings'. However, by interpreting Locke as supplying a semantics instead of an ontology for secondary qualities, Jacovides has Locke supposing that natural philosophy has taught us that there are no colours in bodies at all, that such that colour talk is merely a manner of speaking. But that seems clearly an overreaction to mechanism.

Two other recent commentators also land Locke in anti-realism, but in a more complicated fashion. They analyse secondary qualities as powers (rightly, as we've seen) and then (very plausibly) analyse powers as relations but offer strongly anti-realist interpretations of relations for Locke. Further consideration reveals, I argue, that the consequences of these plausible arguments are problematic enough to merit a search for alternatives.

Walter Ott holds that secondary qualities are powers, according to Locke, and that powers are relations (Ott 2009, 159; 173). Where his interpretation gets more controversial is in his account of Lockean relations. Ott attributes to Locke what he calls 'foundational conceptualism: while relations are fully mind-dependent and have no real being, it remains the case that the mind-independent world provides a foundation (and a justification) for us to form the ideas of relations that we do' (67). (Jacovides (2017, 193 n7) interestingly cites this as congruent with his interpretation.) The mind-dependence comes in because relations depend on an act of comparison, so relations actually exist only when a comparison is being made (Ott 2009, 164). The 'foundational' part of foundational conceptualism is supposed to ameliorate what Ott notes would be the devastating problems with conceptualism full stop:

> I think conceptualism, so stated, cannot be Locke's view. If it were, his overall position would be deeply incoherent: he helps himself to talk of causal relations in his account of how simple ideas represent their objects (II. xxx–xxxii), the linchpin of his epistemology. Our ideas must really be caused by external objects for them to represent what they do, and this must hold regardless of whether there is any mind comparing objects and ideas under the concept of causation. And as we have noted, the qualities that make up our ideas of substances are, for the most part, not true qualities but 'nothing else but so many relations' (II. xxiii. 37), that is, powers.   (Ott 2009, 164)

However, the addition of these foundations does not avert the conclusions that lead to incoherence, since on Ott's interpretation causal relations, including powers, do not obtain or exist unless someone is considering them. Thus, secondary qualities, powers, and all relations are mind-dependent and exist only intermittently, despite the fact that their foundations need not. Note also that secondary qualities such as colours are intermittent in a particularly inapt way: it isn't that the apple is red only when someone has the right kind of idea; rather, the apple is red only when somebody considers the relation between the apple and the mind perceiving it.

Matthew Stuart nicely points out both what is wrong with this interpretation of Lockean relations and one way to avoid the problematic conceptualist element of Ott's interpretation. Stuart (2013, 30) asks what it could mean to say that relations are acts of comparing things, given that intrinsic features are all that are needed to account for the truth of a relational claim such as, 'I am three inches shorter than my son.' He then suggests: 'It is more charitable, but also more plausible, to suppose that when Locke says that "The nature...of Relation, consists in the referring, or comparing two things," he is again being careless about the distinction between ideas and their objects' (29). That is to say, acts of comparison are required to generate our idea of relations, but not relations themselves.

Despite this, Stuart's own account of Lockean relations is also problematic, and Stuart also lands Locke in anti-realism about powers. Stuart notes, rightly and importantly, that Locke often treats our mode-ideas and relation-ideas as being similar, as subject to the same kinds of accounts. Further, Locke often says things that sound rather anti-realist about modes. Stuart, however, concludes that Locke must think that there are modes in the world. He holds, insightfully, that when Locke sounds anti-realist about modes, he is really just making the point that mode-ideas have a different direction of fit from substance ideas. Stuart suggests that many of Locke's seemingly anti-realist remarks about relations can be explained in the same way:

> [Such passages do not] show that Locke thinks that there are no relations. They show that he thinks that the conditions that govern the adequacy and reality of ideas of relations are like those that govern the adequacy and reality of ideas of mixed modes. He thinks that neither ideas of mixed modes nor ideas of relations are made with the purpose of modeling stable and unified items in the world. Neither ideas of mixed modes nor ideas of relations are reckoned defective if there happens to be nothing answering to them.   (25)

Nevertheless, Stuart holds that one particular passage, II.xxv.8, commits Locke to the view that relations simply do not exist: 'This farther may be considered concerning *Relation*, That though it be not contained in the real existence of Things, but something extraneous and superinduced: yet the *Ideas* which relative Words stand for, are often clearer, and more distinct, than of those Substances to which they do belong' (*Essay* II.xxv.8, 322).

I submit that, given the unattractiveness of the view that relations simply do not exist, plus the deep connections that Locke sees between modes and relations,[9] we have sufficient reason to discount II.xxv.8. Further, there is good reason to see Locke as merely flirting in this passage with the thought that relations are less than fully real. For Locke's use of 'though' is easily read as 'even if', that is 'even supposing that', and his real point in that sentence is a claim about *ideas*. Thus, Locke's claim becomes: even if we suppose that

---

[9] Here is one more point: 'parricide' is one of Locke's examples of mixed modes. Stuart would have Locke be a realist about mixed modes but deny the existence of relations. But surely parricide can't exist without some relation existing, such as paternity.

relations aren't contained in the real existence of things, yet still our ideas of relations are often clearer than other ideas. (Not to mention that 'something extraneous and superinduced', which does belong to substances, doesn't sound like something that fails to exist altogether.) Thus, it is a mistake to see II.xxv.8 as requiring anti-realism about relations. Rather, I suggest, we should assimilate relations to modes and then follow Stuart in holding that Locke endorses the existence of modes.

Unlike Ott, Stuart doesn't highlight the fact that powers are relations on his view. Rather, he presents his account of Locke's secondary qualities as a sort of realism, according to which secondary qualities are attributed to bodies as powers, but as 'degenerate powers', that is, actual causings of ideas. Stuart's motivating thought here is that we can explain Locke's inclination to say that porphyry has no colour in the dark as an expression of the ordinary idea that something isn't really fully *able* to do something until all the circumstances are in place that will allow it to actually do that thing. Thus, the powers have degenerated from potentialities to actual causings.

This degenerate-powers interpretation has several associated difficulties. One is just the degeneration itself: Locke's powers (or, at least, Locke's secondary quality powers) become actualities, not potentialities. But this is flatly inconsistent with how Locke defines 'power' in *Of Power* (II.xxi): 'The Mind...considers in one thing the possibility of having any of its simple *Ideas* changed, and in another the possibility of making that change; and so comes by that *Idea* which we call *Power*' (*Essay* II.xxi.1, 233). Power is clearly possibility or potentiality on the official definition, not an actual causing.[10]

Another problematic issue, forthrightly acknowledged by Stuart (2003, 26), is that, as secondary qualities in the objects come and go, so does kind membership. Thus, it's not just that I can make this ring cease to be yellow by closing my eyes, I can also make it cease to be gold.

One further consequence, not noted by Stuart, is that these causings are occurrent relations, and so they don't exist, on his interpretation. That the powers are or include relations is explicit in his earlier paper, 'Locke's Colors': 'To identify colors with powers in the degenerate sense is to identify them with relational but nondispositional properties of objects. On this view, an object is colored just so long as it is standing in a certain causal relation to an

---

[10] Stuart (2013, 140) acknowledges that Locke does not think that all powers are degenerate powers. Nevertheless, it would surely be unsatisfactory if one of Locke's most central uses of his concept of power failed to fit with his official account of the idea of power, and no explanation or notice were given of this fact.

observer of some sort, with a certain result obtaining' (Stuart 2003, 73; cf. Stuart 2013, 140). If powers depend on relations, and relations do not exist, then colours and other secondary qualities do not exist: thus, the degenerate powers interpretation degenerates also from realism to anti-realism.

I hope to have shown that all of these contemporary anti-realist interpretations of Locke's secondary qualities have some clear disadvantages. In the last section of the chapter (4.3), I offer a general argument that Locke must be some sort of realist about powers, and thus secondary qualities. (Though by realism here I mean something fairly minimal: that Locke must hold that powers exist and that they are in objects, though not necessarily solely in objects.)

4.2. Reductionism?

A further issue about the metaphysics of Locke's powers is whether he supposes that they are fully reducible to their causal basis. Both Pasnau and Ott argue that Locke must be a *reductionist* about the powers that are secondary qualities. Pasnau holds that Locke must be a reductionist because the alternative is just too un-Lockean to countenance. If talk of powers were ontologically committing for Locke, powers would be, he says, bare dispositions in this sense—(1) merely conditionally actual, and (2) not causally efficacious, or only derivatively so—but neither Locke nor Boyle could plausibly be held to believe in such things (Pasnau 2011, 519–29).

But both of Pasnau's claims are disputable. As for (1), his claim that an unreduced power must be merely conditionally actual: the essential nature of a *power* is that it *is* a power or ability, and *that* is actual as long as the thing has the power, so powers need not be understood as merely conditionally actual. As for (2), the worry that unreduced powers would not have causal efficacy, I am tempted to say this is some sort of category mistake, for powers aren't supposed to *have* causal efficacy but rather to *be* causal efficacy. Still, there is something right about Pasnau's point that it is the primary qualities that are doing the causing for Locke—or better, that it is substances doing the causing, but that which effects follow depends on their primary qualities. Some substances, because of their primary qualities, can cause in us an idea of redness, which is to say that those substances have the ability or power that is the quality of redness. But I think this is just what Locke should allow: although he sometimes speaks of powers as though they themselves were causally efficacious, in fact powers are manifestations of the causal efficacy of substances. This is to suggest that (2) is not really a problem, once we consider the relation between substances, primary qualities, and their powers. In sum, in neither Pasnau's (1) nor (2) do we find the basis for an argument that powers must be fully reducible to some underlying basis, because (1) is incorrect—unreduced powers are not merely conditionally actual—and (2) is unproblematic—powers aren't supposed to be causally efficacious, for powers don't have powers, rather they are powers.

The question obviously raised by any reductionist interpretation, such as those proposed by Pasnau and Ott, is: to what should powers, for example secondary qualities, be reduced, exactly? Both are somewhat elusive on this question. Pasnau states that powers, for Boyle and Locke, 'are nothing more than a corpuscularian structure embedded in a certain sort of world' (Pasnau 2011, 520). He notes also that powers have a 'relational character' (527). These are reconcilable, he explains, because the nominal approach to powers 'licenses an ecumenical tolerance for different ways of individuating powers (for example, as relative or non-relative to perceivers)' (533). But surely this is too convenient; if powers are to be fully reduced, as Pasnau holds, mustn't they be reduced to something in particular? Pasnau writes, 'On my nominal account...every way of talking is as good as the next, and the only substantive task is to track our ordinary conceptual framework' (533). But if powers aren't being reduced to anything in particular, then it seems that we end up with something like Jacovides's interpretation, according to which Locke denies the existence of secondary qualities, while leaving us with a semantics for talk about them.

Ott, too, wants powers, and so secondary qualities, to be reducible, and he too is somewhat non-committal about how this would go. Powers are supposed to be 'multilaterally reducible' (Ott 2009, 173) to the qualified body in addition to whatever further circumstances of that body feature in the relation that is the power. But how much of the circumstances are to be included? If every relevant item is included, the result is Stuart's degenerate powers view (according to which powers are the actual-n-place relation among all of the items necessary for the production of the relevant sensory idea, and thus, powers become actual causings).[11]

---

[11] Not to mention that, as we saw above, on Ott's view an act of comparison is needed to constitute the relation.

Ott is somewhat cavalier[12] about reduction talk versus supervenience talk, stating that 'if we prefer, we can say that on the reductive view, a power supervenes on the relevant mechanical qualities of the objects it relates: fixing the mechanical qualities of a set of objects will fix their powers' (2009, 147). But this isn't just a matter of preference, for supervenience talk is much less restrictive than reduction talk. To say that secondary qualities supervene on primary ones requires only that secondary qualities never differ without a difference in primary qualities, and of course this requires covariation, not identification. I think it is clear (and to this extent I agree with Pasnau and Ott) that Locke would accept some version of the supervenience claim. This much follows from what I've elsewhere called Locke's essentialism—once you've specified the primary qualities, real essences, and the spatial arrangements, everything else should follow (Downing 2013).[13]

But should we regard the powers that follow as existents? Ott (2009, 141) and Pasnau (2011, 525; 529) think that this must look to Locke like a dreadful error, perhaps a lapse into scholasticism. But why is this so plain? It would, of course, be a mistake to regard the power that follows as a *res*, a thing in the sense of being capable of separate existence. That would be to revive the real qualities for which Descartes and Boyle castigate scholasticism. It would be a dreadful error also, the key error, the error that Boyle and Locke care most about, if we supposed that the power that is, for example, redness in a body resembled my idea of redness. But that we should suppose that particular primary qualities, rightly situated, give rise to a countless host of abilities, including abilities to cause ideas in perceivers, seems exactly like what Boyle and Locke would want; it is what they urge us to notice.[14]

---

[12] This is quite unlike Pasnau (2011, 519), who makes explicit that as he sees it supervenience is not enough, and that 'power' and 'disposition' must be taken in an 'utterly reductive' sense.

[13] In seeing Locke as committed to essentialism, I follow Michael Ayers. What I call 'essentialism' corresponds closely to what Ayers calls 'pure mechanism' (see Ayers 1981, 210; 1991, 2:135, 153, and 190). I neglect one complication here: if minds are not material, then it isn't clear that particular sensory ideas in us will follow from the primary qualities and real essences of substances (plus spatial arrangements), without some choice on God's part.

[14] It is true that Boyle, when emphasizing 'the relative nature of physical qualities', urges us not to assume that all qualities are 'real', 'physical', 'inherent' in the bodies themselves, and 'distinct' from the matter or primary affections (Boyle 1999–2000, 5:310–14; B 23–9). One might think that Boyle thereby rejects any kind of reification of powers, rejects regarding them as existents. I submit, however, that a weaker reading of this attack is quite plausible, according to which Boyle is rejecting scholastic real (separable) qualities and, also, is emphasizing that powers derive from relations, and thus are perfectly intelligible and explicable. But, furthermore, it would be rash to assimilate Locke to Boyle on this point. For Locke is much more interested than Boyle is in locating secondary qualities in bodies (ideas in us, qualities in things), and this plays a crucial role in his systematic epistemology, as emphasized below in Section 4.3.

4.3. The Epistemology of Powers and Implications thereof

Having burrowed this far into Locke's metaphysics, it may be time for a refreshing turn to epistemology: in interpreting Locke's ontology of qualities, we ought to consider the implications of his epistemology of qualities.

We should recollect that powers are extremely important to Locke's system and to his epistemological conclusions.[15] Locke declares unreservedly in *Essay* II.xxxi.2 that '*all our simple* Ideas *are adequate*' (375). And adequate ideas are those 'which perfectly represent those Archetypes, which the Mind supposes them taken from; which it intends them to stand for, and to which it refers them' (II.xxxi.1, 375). Further, the archetypes of simple ideas such as the ideas of redness, roundness, solidity, bitterness (primary and secondary quality ideas alike, notice), are all powers.

Why are these ideas all adequate?

> Because being nothing but the effects of certain Powers in Things, fitted and ordained by GOD, to produce such Sensations in us, they cannot but be correspondent, and adequate to those Powers: And we are sure they agree to the reality of Things. For if Sugar produce in us the *Ideas*, which we call Whiteness and Sweetness, we are sure there is a power in Sugar to produce those *Ideas* in our Minds, or else they could not have been produced by it. And so each Sensation answering the Power, that operates on any of our Senses, the *Idea* so produced, is a real *Idea*, (and not a fiction of the Mind, which has no power to produce any simple *Idea*;) and cannot but be adequate, since it ought only to answer that power: And so all simple *Ideas* are adequate.   (*Essay* II.xxxi.2, 375)

This is a crucial epistemological point, for Locke. For the claim that all our simple ideas are adequate to come out true, it is necessary that our ideas be aimed at powers, that there *be* powers, and that those powers be completely known. Anything less makes the ideas inadequate: '*Inadequate Ideas* are such, which are but a partial, or incomplete representation of those Archetypes to which they are referred' (II.xxxi.1, 375). This requires some sort of realism about powers. Further, it cuts against reductionism, because the purported reduction base is typically unknown, while the powers, Locke holds, are fully known. Powers must remain real, easily known, and readily collected from

---

[15] This is an importance that can be traced back to the early drafts of the *Essay*.

the phenomena in order for our ontology of powers to remain in harmony with Locke's epistemological use of powers. Once we see that secondary qualities are powers barely, we see that they are both in bodies and entirely within our grasp.

References

Ayers, Michael. 1981. 'Mechanism, Superaddition, and the Proof of God's Existence in Locke's *Essay*'. *Philosophical Review* 90 (April): 210–51.

Ayers, Michael. 1991. *Locke: Epistemology and Ontology*. 2 vols. New York: Routledge.

Bayle, Pierre. 1734–8. *The Dictionary Historical and Critical of Mr. Peter Bayle*. 2nd ed. London.

Boyle, Robert. 1999–2000. *The Works of Robert Boyle*. Ed. Michael Hunter and Edward B. Davis. 14 vols. London: Pickering & Chatto.

Downing, Lisa. 1998. 'The Status of Mechanism in Locke's *Essay*'. *Philosophical Review* 107, no. 3 (July): 381–414.

Downing, Lisa. 2011. 'Sensible Qualities and Material Bodies in Descartes and Boyle'. In *Primary and Secondary Qualities: The Historical and Ongoing Debate*, ed. Lawrence Nolan, 109–35. Oxford: Oxford University Press.

Downing, Lisa. 2013. 'Mechanism and Essentialism in Locke's Thought'. In *Debates in Modern Philosophy*, ed. Stewart Duncan and Antonia LoLordo, 159–69. New York: Routledge.

Galilei, Galileo. 2008. *The Essential Galileo*. Ed. and trans. Maurice A. Finocchiaro. Indianapolis: Hackett Publishing Co.

Hobbes, Thomas. 1962. *The English Works of Thomas Hobbes*. Ed. William Molesworth. Darmstadt: Scientia Verlag Aalen.

Jacovides, Michael. 2017. *Locke's Image of the World*. Oxford: Oxford University Press.

Lee, Mi-Kyoung. 2011. 'The Distinction between Primary and Secondary Qualities in Ancient Greek Philosophy'. In *Primary and Secondary Qualities: The Historical and Ongoing Debate*, ed. Lawrence Nolan, 15–40. Oxford: Oxford University Press.

Menn, Stephen. 1995. 'The Greatest Stumbling Block: Descartes' Denial of Real Qualities'. In *Descartes and His Contemporaries: Meditations, Objections, and Replies*, ed. Roger Ariew and Marjore. Grene, 182–207. Chicago: University of Chicago Press.

Ott, Walter. 2009. *Causation and Laws of Nature in Early Modern Philosophy.* Oxford: Oxford University Press.

Pasnau, Robert. 2011. *Metaphysical Themes 1274–1671.* Oxford: Oxford University Press.

Rickless, Samuel. 2014. *Locke.* Oxford: Blackwell.

Schmaltz, Tad M. 1996. *Malebranche's Theory of the Soul: A Cartesian Interpretation.* Oxford: Oxford University Press.

Stuart, Matthew. 2003. 'Locke's Colors'. *The Philosophical Review* 112, no. 1 (January): 57–96.

Stuart, Matthew. 2013. *Locke's Metaphysics.* Oxford: Oxford University Press.

Wilson, Robert A. 2002. 'Locke's Primary Qualities'. *Journal of the History of Philosophy* 40, no. 2 (April): 201–28.

9

Hume on Causation and Causal Powers

Peter Millican

David Hume is perhaps most celebrated for his analysis of causation and of inductive causal reasoning. Moreover, his quest to understand causal power and necessity played a central role in his philosophy and was arguably the primary stimulus behind his *Treatise of Human Nature* (Millican 2016, 86–93). The longest part of that work, Book 1 Part 3, is centred around his analysis of the relation of cause and effect, whose main component is revealed at *T* 1.3.2.11 to be the idea of *necessary connexion*, which he later virtually equates with the idea of causal *power* (*T* 1.3.14.4, 1.3.14.19, 1.3.14.28). After exploring various associated byways into the Causal Maxim, inductive inference, belief, and probability, Hume's quest for the origin of this idea culminates in Section 1.3.14, 'Of the idea of necessary connexion', which finally identifies the crucial *impression* of necessary connexion from which the corresponding *idea* must be copied (in accordance with his Copy Principle).[1] Surprisingly, that impression turns out to be the customary inference of the mind that we characteristically make in response to observed regularities: having repeatedly seen A followed by B—what Hume calls a *constant conjunction* between A and B—we just find ourselves inferring B from A (a psychological propensity Hume calls *custom*), and it is this inferential tendency which leads us to think of A and B as causally connected and to view B as following *necessarily* from A. Essentially the same account is presented in Hume's first *Enquiry*, published in 1748 and

---

[1] The Copy Principle is stated at *T* 1.1.1.7 and its application in the case of necessary connexion is summarized at *T* 1.3.14.1. Crudely, Humean *ideas* are thoughts, *impressions* are sensations or feelings, and the term *perceptions* covers both: 'By the term *impression* . . . I mean all our more lively perceptions, when we hear, or see, or feel, or love, or hate, or desire, or will. And impressions are distinguished from ideas, which are the less lively perceptions, of which we are conscious when we reflect on any of those sensations or movements above mentioned.' *EHU* 2.3.

Peter Millican, *Hume on Causation and Causal Powers* In: *Reconsidering Causal Powers: Historical and Conceptual Perspectives*. Edited by: Benjamin Hill, Henrik Lagerlund, and Stathis Psillos, Oxford University Press (2021).
© Peter Millican.
DOI: 10.1093/oso/9780198869528.003.0010

reissued in many editions throughout his life. Thus so far, at least, we can be reasonably confident that it represents his settled and enduring view.

Having identified the impression of power or necessary connexion as deriving from mental inference rather than external perception, Hume in the *Treatise* repeatedly stresses the apparently paradoxical nature of this result. Causal necessity, he says, 'is nothing but an internal impression of the mind', and 'exists in the mind, not in objects'; again, 'power and necessity...are...qualities of perceptions, not of objects'. Indeed, *we cannot even form an idea of causal power or necessity as a quality of external objects* (*T* 1.3.14.20–7). Understandably, some readers have taken Hume here to be denying that there are any 'objective' causal relations, in the sense of causal relations that can be ascribed truly or falsely, irrespective of the subjective observer's point of view.[2] Such *causal subjectivism*, however, seems hard to square with Hume's clear endorsement of causal science, implicit in his over-all project of an empirical 'science of man' (*T* Intro.4–10). It also seems to be contradicted by the first of his famous two 'definitions of cause' (*T* 1.3.14.31), which defines a cause in terms of objective relations of constant conjunction. Just a few paragraphs later, Hume seems to be even more explicitly objectivist in proposing 'general rules, by which we may know when...objects...*really are*...causes or effects to each other' (*T* 1.3.15.2, my emphasis). These rules are further elaborations of the claim '*that the constant conjunction of objects determines their causation*' (*T* 1.3.15.1, Hume's emphasis), again suggesting that he intends to reduce causation to objective regularity relations and is thus a *reductionist* or *regularity theorist*. Nor is there any plausible comprom-ise in taking Hume to be subjectivist about *causal necessity* but objectivist about *causation*, given both his repeated insistence that necessity is essential to causation, and his later definition of necessity in exactly parallel terms.

Faced with these apparently conflicting strands in Hume's theory of caus-ation, various subtler readings have recently been proposed to reconcile them. *Projectivist* interpretations, for example, view causal thought and language as a *projection* of our natural inferential tendencies. Such projection can lead to mistaken objectification (thus potentially explaining Hume's subjectivist rhetoric at *T* 1.3.14.20–7) but need not imply an *error theory* according to which causal language is irredeemably wrong-headed. Thus, in particular,

[2] This understanding of 'objective' should be borne in mind in what follows, for the word is notori-ously slippery and interpretable in various other ways (e.g., 'in objects', 'non-mental', 'precisely meas-urable', 'unbiased', or even 'intersubjective').

quasi-realist interpretations take Hume's analysis as itself providing a *justifica-tion* of objectivist talk about causal relations and powers, permitting us to consider causal judgements as true (or false) even though such relations and powers do not feature metaphysically as part of the furniture of the universe. At the other extreme—and in far more radical contrast to traditional under-standings of Hume—*sceptical realist* interpretations take causal relations and causal powers to involve *absolute* (*aprioristic*) metaphysical necessities quite independent of human language or judgement, and read Hume's discussion of causation not as *denying* such relations or powers, but rather as questioning our ability to understand or know anything about them. These 'New Hume' interpretations have taken encouragement from Hume's apparently sincere talk of *powers* in the first *Enquiry*, supposedly evincing his commitment to real powers at the fundamental metaphysical level.

In this chapter, I shall try to establish a reliable picture of Hume's view on causation and causal powers, primarily by close analysis of his relevant texts. But rather than diving immediately into complex interpretative debates, I shall start by attempting to establish twelve relatively straightforward *key points* about Hume's theory, all of which can be backed up very strongly and consistently from those texts. Only then shall I turn to the deeper interpret-ative issues, arguing that the key points indicate clearly where Hume's own views are to be found. To summarize, I shall be arguing that Hume is essen-tially a *reductionist* about causation and causal powers (with at most some modest hints of *projectivism*). Reductionism implies that Hume is *objectiv-ist* about causes, in the sense specified earlier: he sees the ascription of causal relations and powers as potentially true or false, irrespective of the subjective observer's point of view. Whether he believes in *causal powers in objects*, how-ever, turns out to be a more delicate question, as we shall see in the final section.

1. Key Points of Hume's Theory of Causation

Let us now examine those Humean tenets about causation whose textual sup-port is sufficiently clear and consistent to justify treating them—at least for the present—as well-established 'key points' of his theory, before going on to consider their implications for his overall metaphysics of causation. These key points might ultimately have to be challenged if we find that they conflict with each other (or with yet further texts), but it seems obviously desirable, if we can, to find an interpretation that maximally respects them.

1.1. Whether A Causes B is an Objective Matter of Fact, and Causes—whether Superficial or Hidden—Can Be Discovered by Systematic Investigation

Hume's investigation of human nature is focused on the *empirical* discovery of *causes*, since only this can ground scientific explanation and inference to the unobserved. His declared aim is to discover the mind's 'powers and qualities... from careful and exact experiments, and the observation of those particular effects, which result from its different circumstances and situations', endeavouring 'to render all our principles as universal as possible, by... explaining all effects from the simplest and fewest causes' (*T* Intro. 9, cf. *EHU* 4.12). As he begins his great investigation, he clearly considers *truth* about these things to be potentially achievable, though it is likely to 'lie very deep and abstruse' and to be discoverable only with great effort and 'pains' (Intro.3).

Much later, having completed his analysis of causation (and as already noted), Hume expands on his first definition of cause by spelling out eight 'Rules by which to judge of causes and effects', whose explicit purpose is to facilitate the empirical discovery of *real* causes: 'Since therefore 'tis possible for all objects to become causes or effects to each other, it may be proper to fix some general rules, by which we may know *when they really are so*'(*T* 1.3.15.1, my emphasis). These rules are intended to help in identifying the genuine causal factors within complex situations, both by distinguishing amongst known factors, but also by prompting deeper investigation to discover factors that are not yet apparent. Such searches for hidden causes are also emphasized elsewhere, notably in a well-known paragraph which highlights how 'philosophers', faced with a 'contrariety of events', often find on further investigation that this superficial variability results from 'the [formerly] secret operation of contrary causes' (*T* 1.3.12.5, repeated at *EHU* 8.13).

1.2. Causes Are Understood to Be Prior and Contiguous to Their Effects

Hume starts his analysis of the idea of *causation* by pointing out that it involves a *relation* between cause and effect (rather than independent qualities of either) and then considering what relational properties might be involved. He quickly identifies *contiguity in time and place* and *temporal priority of the cause*—often abbreviated to 'contiguity and succession'—as obvious candidates and confirms these in his later *rules* (*T* 1.3.2.6–7,

1.3.15.3–4). However, his commitment to the contiguity requirement is not universal since, as he later points out, many of our impressions and ideas have no physical location and hence are not susceptible of spatial contiguity (*T* 1.4.5.10–14, referenced from *T* 1.3.2.6). In the first *Enquiry*, contiguity is not mentioned at all as a requirement on causal relations.[3]

1.3. The Principal Component of the Concept of Causation is Necessary Connexion, which is Essential to it

The most important component of the concept of causation, however, is neither contiguity nor priority of cause to effect: 'Shall we then rest contented with these two relations of contiguity and succession, as affording a compleat idea of causation? By no means. An object may be contiguous and prior to another, without being consider'd as its cause. There is a NECESSARY CONNEXION to be taken into consideration; and that relation is of much greater importance, than any of the other two above-mention'd' (*T* 1.3.2.11). That some particular event A is contiguous and prior to event B does not imply that A is the cause of B: we take causation between them to involve also some sort of connexion, whereby A's occurrence brings B about, or necessitates B. Hume is here saying that necessary connexion is distinct from *single-case* contiguity and succession. A few sections later, when considering causal inference, he will introduce *repeated* events into his discussion, and talk of *constant conjunction*: 'we have...discover'd a new relation betwixt cause and effect,...This relation is their CONSTANT CONJUNCTION. Contiguity and succession are not sufficient to make us pronounce any two objects to be cause and effect, unless we perceive, that these two relations are preserv'd in several instances. We may now see the advantage of quitting the direct survey of this relation, in order to discover the nature of that necessary connexion, which makes so essential a part of it' (*T* 1.3.6.3). The wording and capitalization clearly refer back to *T* 1.3.2.11, signalling that constant conjunction—repeated contiguity and succession—is destined to provide the key to the idea of *necessary connexion*; it will do this by providing the ground of causal inference. Accordingly, the paragraph ends with the prophetic sentence: 'Perhaps

---

[3] Hume's increasing awareness of Newtonian physics, with its gravitational action at a distance, could well have provided another reason for his dropping this requirement and might perhaps have given him pause even about the supposed impossibility of action at a *temporal* distance.

'twill appear in the end, that the necessary connexion depends on the inference, instead of the inference's depending on the necessary connexion.'

Although constant conjunction will indeed prove to be key to what follows, we should note that it never displaces necessary connexion from Hume's account of the idea of the causal relation, for he continues to insist (as in the final sentence of the quotation from *T* 1.3.6.3 above) that necessary connexion is *essential* to that relation.[4] He is particularly emphatic about this when discussing liberty and necessity, repeatedly stating that necessity 'makes an essential part' of causation itself (*T* 2.3.1.18; cf. *EHU* 8.25) and of the definitions of cause (*T* 2.3.2.4; *EHU* 8.27). Within these later discussions, the intimate link between causation and necessary connexion is emphasized even more strongly when Hume explicitly frames two definitions of necessity, reflecting those of cause.

1.4. Causal Necessity is Not the Same as *Conceptual* Necessity

Although Hume's terminology on the matter is not entirely consistent, it is clear that he generally presupposes a fundamental distinction between causal and conceptual modalities. Causal necessity is the main topic of *T* 1.3, and the target of the impression hunt which largely structures that part. Conceptual necessity, which Hume sometimes calls 'metaphysical' or 'absolute' necessity, is a stronger notion, applying to propositions that are intuitively or demonstratively certain—what the first *Enquiry* calls 'relations of ideas'—and is intimately linked with his important Conceivability Principle, *that whatever is conceivable is possible*. This contrast, and its significance for understanding Hume's theory of causation, is clear within his discussion of belief: 'with regard to propositions, that are prov'd by intuition or demonstration...the person, who assents, not only conceives the ideas according to the proposition, but is necessarily determin'd to conceive them in that particular manner....Whatever is absurd is unintelligible; nor is it possible for the imagination to conceive any thing contrary to a demonstration. But...in reasonings from causation, and concerning matters of fact, this absolute

---

[4] Hence those who interpret Hume as denying that causal relations involve necessity are certainly mistaken if 'necessity' here is interpreted in Hume's own sense of the term. No doubt he does deny causal necessity as some other philosophers have supposed it to be; indeed, he denies that such philosophers even have any understanding of what they are trying to suppose (*T* 1.3.14.27, 1.4.7.5; *EHU* 7.29). But he seems to be a firm believer in causal necessity in what he insists is the only legitimate sense of the term.

necessity cannot take place, and the imagination is free to conceive both sides of the question' (*T* 1.3.7.3). Hume repeatedly emphasizes that causal relations are 'matters of fact' and *cannot* be determined by considerations of conceivability, which is why they can be discovered only by experience (rather than through a priori reasoning).[5] Expressed in terms of *conceptual* modality, therefore, 'Any thing may produce any thing. Creation, annihilation, motion, reason, volition; all these may arise from one another, or from any other object we can imagine' (*T* 1.3.15.1, cf. 1.4.5.30, 1.4.5.32). And 'The mind can always *conceive* any effect to follow from any cause, and indeed any event to follow upon another: whatever we *conceive* is possible, at least in a metaphysical sense' (Hume *ABST* 11, cf. *EHU* 12.28–9). Thus, where A causes B, the two will be necessarily connected in a *causal* sense, but not in a *conceptual* (that is, *absolute* or *metaphysical*) sense. Hence it is vital not to confuse these types of modality when considering Hume's philosophy of causation.[6]

1.5. Hume is a Convinced Determinist, although his Basis for this is Unclear

Hume is a *determinist*, in the sense that he believes the course of events to be completely determined by antecedent conditions and temporally uniform causal laws. This thesis is compatible with a wide range of theories about what causation involves, thus allowing considerable variation in types of determinism. Presumably for Hume himself, however, it ultimately comes down to all events in the universe occurring *in conformity with* the relevant laws. So it does not require, for example, any 'deeper' metaphysical necessity underlying either the specific laws or the law-governed nature of the universe as a whole. Indeed, on Humean principles we cannot possibly expect any such 'deeper' explanation of determinism, given that it is a matter of fact rather than a conceptual truth.

---

[5] See, for example: Hume *ABST* 18, 21; *EHU* 4.4, 4.14, 4.19, 5.3, 5.20, 7.27, 7.29, 12.28. 'Hume's Fork' between relations of ideas and matters of fact is introduced in *EHU* 4.1–2 and provides a significant improvement on the theory of relations of the *Treatise*, though the two are similar in spirit. For much more on these matters, see Millican 2017.

[6] As pointed out in Millican 2017 (34), there is particular potential for confusion here because Hume's references to 'possibility' are most often to *conceptual* possibility, whereas his references to 'necessity' (especially in the parts of his works that concern his philosophy of induction and causation) are most often to *causal* necessity. Moreover, the two types of modality can often be mixed, as when we pursue the (absolute) logical consequences of what we take to be (causally) necessary laws, for example in applied mathematics (*EHU* 4.13), or when considering the implications of Hume's Copy Principle (*T* 1.2.6.8, 1.3.14.6, 1.3.14.22, 1.4.5.19–21, 1.4.6.2; *EHU* 7.8).

Hume's determinism has been apparent to most of his interpreters, and rarely questioned.[7] It is most evident in his discussions of liberty and necessity (*T* 2.3.1–2; *EHU* 8) and his denials of *chance* or *indifference* (*T* 1.3.12.1, 2.3.1.18; *EHU* 6.1, 8.25),[8] and is at least strongly suggested by his rules by which to judge of causes and effects (for example *T* 1.3.15.6). Determinism also features in his theological discussions, notably regarding the problem of evil (*EHU* 8.32–36) and the morality of suicide (*Essays* 580). Some of Hume's contemporaries were misled by his discussion of the Causal Maxim—'that *whatever begins to exist, must have a cause of existence*' (*T* 1.3.3.1)—to suppose that he denied it and therefore believed that things could come about without any cause. But as Hume emphasized in his 1745 *Letter from a Gentleman* (*LFG* 26) and a letter to John Stewart of February 1754 (Hume 1932, 1:186), his aim here is not to deny the Causal Maxim, but only to show that it 'is neither intuitively nor demonstrably certain' (*T* 1.3.3.3).

The basis for Hume's determinism is unclear, and may simply exemplify the typical optimism of someone dedicated to the scientific search for hidden causes. The nearest he comes to justifying it is at *EHU* 8.13, where he suggests that such causes are usually to be found. In the subsequent two paragraphs of the *Enquiry*, he backs this up by arguing that, in both the physical and human worlds, 'the irregular events, which outwardly discover themselves, can be no proof, that the laws of nature are not observed with the greatest regularity in its internal operations and government....The internal principles and motives may operate in a uniform manner, notwithstanding these seeming irregularities' (*EHU* 8.14–15).

1.6. *Necessary Connexion* is One of a Family of 'Power' Terms, which Hume Treats as Virtually Synonymous in this Context

Hume is surely right to say that we think of a cause as being related to its effect by more than just contiguity and priority, but we might be less convinced that the missing component is correctly described as *necessary connexion*. In mechanical interactions, for example, we think of one billiard ball that strikes another as communicating some impulse, force, or energy to the

---

[7] Harris (2003; 2005) is the most notable recent exception. Millican (2010) comprehensively answers Harris's contention that Hume is not a determinist, documenting and discussing all the points that are summarized here.

[8] In the *Treatise*, Hume usually interprets 'liberty' to mean indifference (*T* 2.3.1.18, 2.3.2.2, 2.3.2.6–8) and denies its existence (2.3.1.15, 2.3.2.1).

other, which influences its subsequent motion, but we would not say that the impact literally *necessitates* that subsequent motion, which will depend also on other factors in the situation (for example, whether other balls are simultaneously impacting, or whether there is a fixed barrier, glue, or some other impediment to motion). Hume suggests that unscientific people—'the vulgar'—may also disagree that causes necessitate, because they believe in what we might call 'chancy' causation, 'attribut[ing] the uncertainty of events to such an uncertainty in the causes as makes the latter often fail of their usual influence; though they meet with no impediment in their operation' (*T* 1.3.12.5; *EHU* 8.13). Such people may be misguided, as Hume himself goes on to argue, but the very possibility of such a belief makes it implausible to claim that our ordinary concept of *causation* includes literal *necessity* as an essential component, despite key point 1.3 above.

Hume himself, however, treats this as a mere terminological inconvenience, insisting rather glibly that his analysis will apply to the entire family of relevant terms:[9] 'I begin with observing that the terms of *efficacy, agency, power, force, energy, necessity, connexion*, and *productive quality*, are all nearly synonimous' (*T* 1.3.14.4). But this seems implausible. As already noted, power, force, and energy do not suggest the inexorability of necessity. And connexion seems directionally symmetrical (that is if A is connected to B, then B is connected to A), whereas the other terms are not (for example, A can have a power to produce B, without B having the reciprocal power). So Hume here seems to be conflating what are in fact rather different ideas.

1.7. Understanding these Terms Involves Having a Certain Simple *Idea*, which is Copied from a Corresponding *Impression* of Reflection

As we shall see later, the conflation just noted may be related to another potential problem in Hume's account, namely his apparent assumption that the idea in question is *simple*, which is at least suggested by the *Treatise*

---

[9] Likewise, *EHU* 7.3 implies that the 'ideas... of *power, force, energy*, or *necessary connexion*' are all subject to a similar analysis. This identification is fully borne out by the subsequent text, which frequently alternates between the relevant terms, including 'power or necessary connexion' (7.5–6, 7.9, 7.26, 7.28, 7.30), 'connexion or power' (7.26), 'power or energy' (7.7–9, 7.11, 7.15–16, 7.19), 'energy or power' (7.10), 'power or force' (7.8, 7.12, 7.21), 'force or power' (7.26), and 'force or energy' (7.16, 7.21, 7.25). Later, Hume (*EHU* 8.25 n19; cf. *T* 1.3.2.10) points out that *producing* is another term in the same family, as is *by which*.

discussion and is explicitly stated in the *Enquiry*.[10] But if there is any such thing as 'the idea of power or necessary connexion',[11] it is hard to see how it can possibly be a simple idea. For necessary connexion is clearly a relation (*T* 1.3.2.6, 1.3.2.11), and relations seem inevitably complex, as Hume himself acknowledges (*T* 1.1.4.7). Likewise, power must be understood, he says, as relative to an effect (*EHU* 7.14, 7.29 n17), so it is hard to see how an impression or idea of *power* can be simple either.

We shall return to these problems in the interpretative discussion below. But for now, let us note that Hume does in fact treat the idea of 'power or necessary connexion' as simple and that he considers it to be derived from an 'internal impression, or impression of reflection' (*T* 1.3.14.22), 'which we *feel* in the mind' (*EHU* 7.29; cf. *EHU* 7.30; *T* 1.3.14.20, 1.3.14.28–9) when we make an inductive inference. The nature of this impression is not entirely clear, however. For although Hume refers to it as something *felt*, most often in the *Treatise* he calls it a 'determination' of the mind or thought, and in the *Enquiry* a 'customary transition', neither of which sounds like a genuine feeling. Again, we shall leave this tricky interpretative issue for later, since the texts by themselves do not give a clear verdict.

1.8. That Impression Arises from Observed Constant Conjunction and the Consequent Tendency to Draw Inductive Inferences

Whatever Hume's views might be on the precise nature of the impression of power or necessary connexion, he is quite clear about the circumstances that give rise to it, namely, repeated observations of a constant conjunction between A and B, followed by a specific observation of just one of the pair. In these circumstances, we find ourselves irresistibly expecting (or drawing an inference to) the other of the pair though a process that Hume calls *custom*. This harks back to his discussion of induction, and his prophetic comment from *T* 1.3.6.3: 'Perhaps 'twill appear in the end, that the necessary connexion

---

[10] *EHU* 7.8 n12 presumes that 'the idea of power' is an 'original, simple idea' in the course of criticising Locke's account of its origin (see also *T* 1.3.14.5). The simplicity of the idea is also strongly suggested by Hume's apparent denial that it can be defined (*T* 1.3.14.4; *EHU* 7.4–5), as complex ideas may be.

[11] See note 9 for paragraphs in the *Enquiry* that use this phrase, which also occurred in the original title of *Enquiry* 7, with the words 'power or' being deleted from the Third Edition (1756) onwards. In the *Treatise*, the precise phrase does not occur, but there is one reference to 'the idea of power or necessity' (*T* 1.3.14.19).

depends on the inference, instead of the inference's depending on the necessary connexion.' The instinctive inference comes first, and explains our ascription of causal necessity, whereas naïvely we might expect that inductive inference would depend on prior causal beliefs.

Once we have explicitly ascribed a causal connexion between A and B, however, this order of explanation changes, and we can then go on to make further inferences—often of much greater complexity—based on that ascription, beyond the simple, instinctive cases of customary inference which Hume initially discusses. And even in the ascription of causes, careful reflective reasoning is often required to distinguish genuine causal relations from those that are merely superficial, as Hume emphasizes when discussing 'unphilosophical probability' and his rules by which to judge of causes and effects (*T* 1.3.13, 1.3.15). So, his initial simple story 'of the idea of necessary connexion' (*T* 1.3.14; *EHU* 7) is apparently intended to focus on the origin of that idea within simple customary inference, and does not pretend to cover its application in general.

1.9. Hume Accordingly Provides Two Definitions of Cause

The famous two 'definitions of cause', which come at the culmination of Hume's account 'Of the idea of necessary connexion' in both the *Treatise* and the *Enquiry*, apparently aim to capture the two crucial circumstances from which that idea arises: *constant conjunction* and *inference of the mind*. Hume writes:[12]

> We may define a cause to be [1] 'An object precedent and contiguous to another, and where all the objects resembling the former are plac'd in like relations of precedency and contiguity to those objects, that resemble the latter.' If this definition be esteem'd defective, because drawn from objects foreign to the cause,[13] we may substitute this other definition in its place,

---

[12] I have inserted numbers into the quotations below to signpost the first and second definitions. Note that the *Enquiry* contains a counterfactual variation on the first definition, marked as 1c, which is not equivalent to 1 because 1 specifies an implication from the first 'object' to the second, whereas 1c specifies an implication from *absence of* the first to *absence of* the second. In the *Treatise*, the fourth rule by which to judge of causes and effects appears to make a similar conflation: 'The same cause always produces the same effect, and the same effect never arises but from the same cause' (*T* 1.3.15.6).

[13] The definition is 'drawn from objects foreign to the cause' because on Hume's account whether some particular 'object' is a cause depends on patterns of behaviour amongst other 'objects' and so is not dependent purely on the particular instance. Hume makes no such deprecatory comment about

viz. [2] 'A cause is an object precedent and contiguous to another, and so united with it, that the idea of the one determines the mind to form the idea of the other, and the impression of the one to form a more lively idea of the other.' (*T* 1.3.14.31)

Similar objects are always conjoined with similar. Of this we have experience. Suitably to this experience, therefore, we may define a cause to be [1] *an object, followed by another, and where all the objects, similar to the first, are followed by objects similar to the second.* Or in other words, [1c] *where, if the first object had not been, the second never had existed.* The appearance of a cause always conveys the mind, by a customary transition, to the idea of the effect. Of this also we have experience. We may, therefore, suitably to this experience, form another definition of cause; and call it, [2] *an object followed by another, and whose appearance always conveys the thought to that other.* (*EHU* 7.29)

There has been much scholarly debate over how these 'definitions' should be interpreted.[14] But it seems unlikely, in view of the points made above, that Hume's intention here is to provide two distinct analytical specifications of the necessary and sufficient conditions for one thing to be the cause of another. He is well aware that customary inference can occur in respect of conjunctions that are very far from constant, as, for example, with the prejudices mentioned at *T* 1.3.13.7. He is also well aware that genuine constant conjunctions can lie undiscovered, so that 'philosophers' who wish to identify them have to go to great trouble to do so (*T* 1.3.12.5, 1.3.15; *EHU* 8.13). So the two definitions—one couched in terms of constant conjunction, and the other in terms of customary inference—will often come apart in practice. This need not be seen as a major problem, however, if we set the definitions in their appropriate context of Hume's theory of ideas rather than anachronistically expecting him to be engaged in an analytic investigation of the type that would become popular more than two centuries later (for example Mackie 1965; 1974).[15] For Hume's primary aim seems to be to investigate our understanding of the relation of cause and effect in terms of [1] the circumstances

his definitions of *necessity* (discussed in Section 1.10 below), perhaps because necessity is standardly understood to be a universal relation rather than a property of an individual object or pair of objects.

[14] For a useful overview see Garrett (1997, 97–101). For a discussion comparing Garrett's approach to the one favoured here, see Millican 2009, 659–66.

[15] The nearest Hume comes to anything that bears comparison with such an analysis is in his 'Rules by which to judge of causes and effects' (*T* 1.3.15).

in which the idea of causal necessity arises, and [2] the impression from which the relevant idea is copied. The two definitions thus aim to sum up these two distinct aspects of his lengthy investigation into the origin of the idea of cause, and in particular its most important component, the idea of necessary connexion.

1.10. Hume Also Provides Two Definitions of Necessity, which He Applies to the Issue of 'Liberty and Necessity'

The most important application of Hume's investigation of the idea of necessary connexion is to the topic of 'liberty and necessity' (roughly, what we would now call free will and determinism).[16] Here, however, Hume's focus is not so much on identifying *causes* as identifying *causal necessity*. Hence he does not directly apply his two 'definitions of cause', but instead applies two corresponding definitions of *necessity*: '*Necessity* may be defined two ways, conformably to the two definitions of *cause*, of which it makes an essential part. It consists either in *the constant conjunction of like objects, or in the inference of the understanding from one object to another*' (*EHU* 8.27; see also *T* 2.3.2.4). Hume considers the distinction between the two pairs of definitions sufficiently important that in his index to the *Enquiry* (*EHU* 307–10) he gives separate entries for 'CAUSE and EFFECT ... Its Definition', referring to 7.29 and 8.25 n19, and for 'NECESSITY, its definition', referring to 8.5 and 8.27.

Hume wields these two definitions of necessity to argue that the very same kind of necessity applies both to physical events (such as impacts of billiard balls) and to human actions. The structure—and much of the wording—of this argument is virtually identical in the *Treatise* and the *Enquiry* (see Millican 2007a, 190–3; 2009, 693–702). But it is put most pithily in the *Abstract*, which in part quotes from *T* 2.3.1.4:

'Here then are two particulars, which we are to regard as essential to *necessity, viz.* the constant *union* and the *inference* of the mind; and wherever we discover these we must acknowledge a necessity.' Now nothing is more evident than the constant union of particular actions with particular

---

[16] Another important application comes in 'Of the Immateriality of the Soul' (*T* 1.4.5.29–33), where Hume appeals to his analysis of causation to refute the anti-materialist claim that 'matter and motion' cannot cause thought.

motives.... And...the inference from the one to the other is often as certain as any reasoning concerning bodies:...Our author pretends, that this reasoning puts the whole controversy in a new light, by giving a new definition of necessity. And, indeed, the most zealous advocates for free-will must allow this union and inference with regard to human actions. They will only deny, that this makes the whole of necessity. But then they must shew, that we have an idea of something else in the actions of matter; which, according to the foregoing reasoning, is impossible.   (*ABST* 32–4)

The crucial application of the definitions is in the final sentence, highlighting the impossibility of forming any other notion of necessity beyond Hume's two definitions. Hence those who attempt to draw a distinction between *moral* and *physical* necessity—most famously Samuel Clarke—are refuted: they cannot even form a coherent idea of the 'something else' that they wish to attribute to 'the actions of matter', and hence their would-be distinction cannot even get off the ground.[17]

1.11. When the Two Definitions Come Apart, *Constant Conjunction* Dominates

As pointed out above, Hume's two definitions frequently come apart in practice, since our inferential tendencies do not always correspond with genuine constant conjunctions. But he is clear that, in our causal reasonings, we should attempt to refine those inferential tendencies so that they do thus correspond as far as possible. Hence, for example, we should endeavour to discover the reliable causal conjunctions that underlie superficial inconsistencies (*T* 1.3.12.5; *EHU* 8.13–15), to identify high-level general rules that can overcome our natural prejudices (*T* 1.3.13.11–12), and—more specifically—to

---

[17] Clarke defends the distinction in his *Remarks* (1717, 15–18). Hume's first attack on it is at *T* 1.3.14.33, just two paragraphs after the presentation of his two definitions, strongly confirming that this provided significant motivation for his analysis of causation. Despite the prominence and repetition of Hume's argument against the distinction, however, it is philosophically unpersuasive (except, perhaps, as interpreted *ad hominem* against Clarke and others). For there is available a more plausible way of distinguishing between intentional and physical causation, based not on supposed different types of necessity but instead on the distinction between causal processes that are directed towards some outcome through means-end reasoning and those that are simply the working out of purpose-less laws. Even putting mentality aside, the example of a chess computer illustrates how a physical causal substrate can implement teleological processing that is responsive to relevant rules and goals, thus potentially permitting two quite distinct patterns of causal explanation of the same behaviour. Hume does not consider the possibility of distinguishing between types of explanation rather than types of necessity, and thereby leaves a potential weakness in his overall theory of causation.

apply the rules by which to judge of causes and effects which Hume spells out in *T* 1.3.15 'to distinguish the accidental circumstances from the efficacious causes',[18] something that he apparently considers to be necessary if we are to discover the true causal relationships within *any* 'phaenomenon in nature' (*T* 1.3.15.11). When we are unable to identify genuinely constant causal relationships, moreover, he enjoins us to do the next best thing by reasoning probabilistically, conditioning our expectations by the experienced frequencies (*EHU* 6, 10.3–4; *T* 1.3.11–12). In all of this, Hume is implicitly giving his first definition of cause priority over the second, favouring actual constant conjunctions over our natural inferential tendencies.[19] Another clear example of this priority comes in his discussion 'of the immateriality of the soul', which states boldly 'that all objects, which are found to be constantly conjoin'd, are upon that account only to be regarded as causes and effects' (*T* 1.4.5.32). In other words, constant conjunctions alone are sufficient to determine causal relations, whether or not they happen to correspond with natural human inferential tendencies.

All this tends to confirm again that the point of Hume's second 'definition' is to characterize the original *impression* from which the idea of power or necessary connexion is copied, and it is not intended to provide a general *criterion* for application of that idea. Our natural inferential tendencies are crucial to get us started in inductive inference, but Hume never suggests that— once started—we should accord those natural tendencies (as encapsulated in the second definition) authority over the disciplined observation of constant conjunctions and judgements of probabilities. On the contrary, he frequently emphasizes the priority of the first definition by stating explicitly that 'the very essence' of power, cause and effect, or necessity is constituted by the 'constant conjunction of objects' (*T* 1.4.5.33), the 'multiplicity of resembling instances' (*T* 1.3.14.16), 'constancy' (*EHU* 8.25 n19), or 'uniformity' (*T* 2.3.1.10).

---

[18] At *T* 1.3.13.11, in his discussion of prejudice, Hume gives a footnote reference to *T* 1.3.15, while saying, 'We shall afterwards [note: Sect. 15] take notice of some general rules, by which we ought to regulate our judgment concerning causes and effects;... By them we learn to distinguish the accidental circumstances from the efficacious causes.'

[19] Garrett (1997, 108–13) argues for an *idealized* understanding of the two definitions, whereby we consider the second definition as concerned with the inferential tendencies that we *would* have if fully informed and rational, thus making the two definitions coextensive. But Hume's text exhibits no such apparent sophistication, and he seems to see the role of the second definition as being to identify the impression in question rather than specifying idealized conditions for the application of the resulting idea. Hume recognizes that we need to apply careful, conscious discipline to infer well, rather than just relying on our natural tendencies, and he sees such discipline as often requiring explicit understanding of the first definition and its refinement in his rules. By contrast, being told to infer 'as an idealized reasoner would do' by itself provides no such specific understanding, and hence the second definition—if taken to have the intention of identifying actual causal relations—would be useless to us.

Such consistent patterns could in principle provide a perfectly informed observer with a reliable basis for inductive inference, but it is clearly the uniformities themselves, not the actual occurrence of any inference based on them, that constitute 'the very essence' of causal necessity (as we shall see further in our discussion of *T* 2.3.2.2 and *EHU* 8.22 n18 in Section 2.1 below).

1.12. In the First *Enquiry*, Hume Recognizes More Sophisticated Causal Relations than in the *Treatise*, Mediated by Quantitative Powers and Forces

There is a significant difference between Hume's discussions of causation in the *Treatise* and in the first *Enquiry*, apparently reflecting a more sophisticated understanding of science. In the *Treatise*, Hume seems to be thinking of causal relations as holding almost exclusively between discrete types of event. Thus, he repeatedly talks of 'constant conjunction', and even the rules by which to judge of causes and effects are mostly couched in terms of the absolute presence or absence of particular causal factors: '5.... where several different objects produce the same effect, it must be by means of some quality, which we discover to be common amongst them.... 6.... The difference in the effects of two resembling objects must proceed from that particular, in which they differ' (*T* 1.3.15.7–8). Only Hume's seventh rule gives any hint that we might be dealing with varying quantities, which cannot therefore be characterized in terms of discrete types of 'object': '7. When any object encreases or diminishes with the encrease or diminution of its cause, 'tis to be regarded as a compounded effect, deriv'd from the union of the several different effects, which arise from the several different parts of the cause' (*T* 1.3.15.9). But such talk of 'parts' of the cause still seems unsophisticated and inadequate when compared with the Newtonian physical science of the time, which would treat the impact of billiard balls, for example, not in the crude terms of 'motion in the first ball' causing 'motion in the second' but rather in terms of the relevant velocities, masses, angles, coefficient of restitution, and so forth—all of these being arithmetically quantifiable on a continuous scale.[20]

The most important Newtonian law applicable to the collision of billiard balls is that of the Conservation of Momentum, which makes no obvious

---

[20] Recognition of this crudity in his rules by which to judge of causes and effects might well account for Hume's omission of them from the *Enquiry*. The nearest equivalent in the *Enquiry* is the long note to *EHU* 9.5, in the section 'Of the reason of animals', which could have included such rules had Hume still considered them adequate.

appearance in the *Treatise* but is explicitly mentioned in the *Enquiry*: 'it is a law of motion, discovered by experience, that the moment or force of any body in motion is in the compound ratio or proportion of its solid contents and its velocity' (*EHU* 4.13). This passage shows an awareness that mechanical causation involves arithmetic quantities such as 'moment or force', and three paragraphs later Hume talks again of 'that wonderful force or power, which would carry on a moving body for ever in a continued change of place, and which bodies never lose but by communicating it to others' (*EHU* 4.16). 'Secret powers' are accordingly mentioned repeatedly both at *EHU* 4.16 and *EHU* 4.21, though Hume's discussion is still couched in terms of constant conjunctions (but now between 'sensible qualities' and 'secret powers' rather than between 'objects' as in the *Treatise*). In a footnote to *EHU* 4.16, however, Hume points out that his talk of powers here is 'loose and popular', referring forward to the 'more accurate explication' that will be delivered in *EHU* 7.[21]

Even in Section 7 of the *Enquiry*, however, the main text shows only modest evidence of Hume's increasing awareness that physical causation involves continuously varying quantities rather than discrete types of event, and only in two notes does this come through relatively clearly:

> We find by experience, that a body at rest or in motion continues for ever in its present state, till put from it by some new cause; and that a body impelled takes as much motion from the impelling body as it acquires itself. These are facts. When we call this a *vis inertiae*, we only mark these facts, without pretending to have any idea of the inert power; in the same manner as, when we talk of gravity, we mean certain effects, without comprehending that active power.   (*EHU* 7.25 n16)

> According to these explications and definitions, the idea of *power* is relative as much as that of cause; and both have a reference to an effect, or some other event constantly conjoined with the former. When we consider the *unknown* circumstance of an object, by which the degree or quantity of its effect is fixed and determined, we call that its power: And accordingly, it is allowed by all philosophers, that the effect is the measure of the power. But if they had any idea of power, as it is in itself, why could not they measure it in itself? The dispute whether the force of a body in motion be as its velocity, or

---

[21] The main point of this footnote (*EHU* 4.16 n7) seems to have been to counter an objection made by Henry Home (Kames 2005, 188–9), who considered the references to powers in the first three paragraphs of *EHU* 4.16 to be inconsistent with the theory of causation that Hume would later present in Section 7. See Millican 2007b, 236–7.

the square of its velocity; this dispute, I say, needed not be decided by comparing its effects in equal or unequal times; but by a direct mensuration and comparison. (*EHU* 7.29 n17)

These notes seem to be intended to bring quantitative 'powers' within the scope of Hume's theory of causation, generalizing beyond mere constant conjunction—and even beyond the multifactor interactions envisaged by the rules of the *Treatise*—to include arithmetical functional relationships. But Hume's main point here seems to be his insistence that our only grasp of such powers is relational: we have no idea or comprehension of them as they are in themselves, and no means of assessing them except in terms of their perceived effects. It is helpful to have this confirmation that Hume intended his language of powers to fit within his 'official' overall theory of causation, but these notes provide at best a sketch of how that theory would be refined to accommodate them.

2. Reductionism, Subjectivism, and Projectivism

The key points itemized above are clearly supported by multiple Humean texts and clearly contradicted by none; hence they should be relatively uncontroversial. Taken together, they strongly support a traditional *reductionist* reading of Hume's theory of causation, with causal relations being objective, universal, and determined by constant conjunctions (or functional relationships of a more complex sort). Yet in recent years, many scholars have rejected this traditional style of interpretation, preferring instead to see Hume as either a *projectivist* or a *sceptical realist*.[22]

The most influential objections to the reductionist reading have focused on what many readers are likely to consider to be the most conspicuous omission from my key points above, namely Hume's notorious and emphatic declarations in *T* 1.3.14 to the effect that 'necessity is...in the mind, not in objects'

---

[22] To summarize several recent overviews of the interpretative landscape: Dauer (2008, 94–8) takes the three main types of reading to be 'reductionist', 'realist', and 'intermediate', the last exemplified by Blackburn's 'projectivism'. Beebee (2012, 137–43) follows the same order but with different labels, dividing the potential interpretations into 'traditional', 'sceptical realist', and 'projectivist'. Beebee (2016, 235–43) likewise starts with the traditional 'regularity theory', but then goes on to 'projectivist' theories before turning to 'sceptical realism'. Garrett (2015, 82–99) uses different terms but agrees that the three main options are 'causal projectivism', 'causal reductionism', and 'causal realism'. His own view, however, combines elements of all three, together with the novel idea that Hume has a 'causal-sense theory' (for comments on which, see Millican 2014, 216–19).

(*T* 1.3.14.22).[23] Such declarations sit uneasily with reductionism, for if causal necessity were indeed a matter of constant conjunction, then it ought to be as objective as the constant conjunctions themselves rather than mind-dependent (cf. *T* 1.3.14.28, where Hume acknowledges that objects' regular 'relations of contiguity and succession' are 'independent of, and antecedent to the operations of the understanding'). So if, contrariwise, we honour Hume's subjectivist declarations and place necessity only 'in the mind', this seems to imply a rejection of reductionism.

Below I shall explain why I consider Hume's subjectivist declarations to be undeserving of the interpretative weight they have generally been accorded. In short, they are overblown, prominent only in the youthful *Treatise*, in tension with Hume's more consistent commitments (notably his first definition of necessity), and therefore best understood not as considered conclusions about causation's metaphysical status, but rather as dramatic expressions of Hume's surprising result regarding the origin of our idea of causal necessity. He is pushed towards subjectivism by his identification of the key 'impression' as something felt 'in the mind', for it is hard to see how an idea copied from an internal feeling could represent something existing outside a mind. Though far less emphasized in the *Enquiry*, this source of tension with objectivist reductionism remains there, but below I shall propose a way of understanding Hume's theory that makes reasonable sense of what he says while also dealing with two other internal problems that were noted above (in Sections 1.6 and 1.7). These tensions in his texts, I shall argue, arise not from any serious doubt on his part about the objectivity of causal relations but, rather, from an impoverished view of reflection, which is evident in other contexts and hinders him from expressing his theory in the most consistent manner. Understood in this light, Hume's theory remains fundamentally reductionist and objectivist, and these conclusions are not threatened by the element of projectivism implicit in his empiricist account of our causal thinking.

2.1. Is Humean Necessity Only 'In the Mind'?

There is an obvious tension between Hume's apparent belief in objective causal relations (key points 1.1, 1.5, 1.11, and 1.12) and his pronouncements—forcefully and conspicuously repeated in the *Treatise*—that the

---

[23] Beebee (2006, 216; 2012, 137–8) and Dauer (2008, 95), for example, both see the crucial objection to reductionist interpretations as deriving from the mind-dependence of Humean necessity.

power, necessity, efficacy, or energy of causes are in the mind, not in objects (T 1.3.14.20, 1.3.14.22–4, 1.4.7.5, 2.3.1.4, 2.3.1.6) and that we cannot even form an idea of them as qualities of objects (T 1.3.14.22, 1.3.14.25, 1.3.14.27, 1.4.3.9, 1.4.7.5). Hume apparently sees these claims as following straightforwardly from his identification of the impression of power or necessity as the 'determination of the mind', and his view—likewise frequently reiterated in the *Treatise*—that ideas can only represent the impression from which they are copied (T 1.1.1.7, 1.1.1.12, 1.2.3.4, 1.2.3.11, 1.3.7.5, 1.3.14.6, 1.3.14.11, 1.4.5.21). But such strident subjectivism sits very uneasily with his two definitions of necessity, reflecting only the second and apparently conflicting with the first. For if causal necessity can indeed be defined in terms of objective constant conjunction, then why can we not frame thoughts about it in those terms also?

Rather than attempting to square all of Hume's problematic statements in the *Treatise*, we can conveniently sidestep them by observing that the presentation of his theory in the later *Enquiry* changes enormously in the relevant respects. There he never says that ideas can only represent impressions, and only once does he even get close to saying that necessity is in the mind, not in objects:

> The necessity of any action, whether of matter or of mind, is not, properly speaking, a quality in the agent, but in any thinking or intelligent being, who may consider the action; and it consists chiefly in the determination of his thoughts to infer the existence of that action from some preceding objects; as liberty, when opposed to necessity, is nothing but the want of that determination, and a certain looseness or indifference, which we feel, in passing, or not passing, from the idea of one object to that of any succeeding one.
>
> (*EHU* 8.22 n18)

Moreover, this single passage is of limited significance. It is part of a note copied largely verbatim from the text of T 2.3.2.2, whose point—within Hume's discussion of liberty and necessity—is to explain 'The prevalence of the doctrine of liberty' in terms of 'a false sensation or seeming experience…of liberty or indifference'. Here the pertinent contrast is between the determination of our thoughts that applies in the case of necessity, and the apparent 'want of that determination…which we feel' in other cases. The note ends by pointing out 'that, however we may imagine we feel a liberty within ourselves, a spectator can commonly infer our actions from our motives and character; and even where he cannot, he concludes in general, that he might, were he perfectly acquainted with every circumstance of our situation and temper, and the most secret springs of our complexion and disposition. Now this is the very essence of necessity, according to the foregoing doctrine.' Here 'the very essence of necessity' turns out to be the *potential* for well-informed inference—based on the relevant constant conjunctions—rather than *actual* inference. It thus corrects any misleading impression that the wording earlier in the note might give towards the sort of extreme subjectivism that was so conspicuous in the *Treatise*.[24]

There is just one other passage in the *Enquiry* that might be thought to imply such extreme subjectivism, and this occurs in a far more significant location, namely the paragraph prior to the two definitions in Hume's discussion 'Of the idea of necessary connexion': 'When we say, therefore, that one object is connected with another, we mean only, that they have acquired a connexion in our thought, and give rise to this inference, by which they become proofs of each other's existence' (*EHU* 7.28). But this too turns out to be much less than it appears. First, the focus of this paragraph is very clearly on the *origin* of the relevant idea, and on how we come to say that 'objects' are connected. Secondly, just one paragraph later—having given his two definitions—Hume goes on to state very explicitly that we can indeed 'mean' something more by our attributions of causal connexion, echoing now the first definition as well as the second: 'We say, for instance, that the vibration of this string is the cause of this particular sound. But what do we mean by that affirmation? We either mean, *that this vibration is followed by this sound, and that all similar vibrations have been followed by similar sounds: Or, that this vibration is followed by this sound, and that upon the appearance of one, the mind anticipates the senses, and forms immediately an idea of the other*' (*EHU* 7.29). The proximity of these two passages seems unlikely to be coincidental,[25] suggesting that the latter is intended as a deliberate extension or correction of the former. When we first infer from observed A to anticipated B by custom, this naturally leads us to assert a connexion between them, and at that stage 'we can only legitimately mean' that they are connected in our thought.[26] Having gone on to analyse the objective circumstances that generate this

---

[24] Note also that the earlier wording—'a quality...in any thinking or intelligent being, who *may* consider the action'—is naturally readable as expressing a potentially *counterfactual* conditional, rather than applying only where the 'determination of the mind' actually takes place.

[25] The phrase 'we mean' occurs in only twelve paragraphs of Hume's philosophical works, and this is the only case of its occurring in adjacent paragraphs.

[26] The scare quotes here highlight that phrases like 'we only mean X' are typically less than rigorous and often used loosely. Hume twice in the *Treatise* uses the phrase 'mean nothing but X', once in connexion with the will and once as applied to moral pronouncements (*T* 2.3.1.2, 3.1.1.26), and neither of these occurrences seems to be meant literally.

connexion, however, and after framing Hume's two definitions, we can then mean something more, namely the obtaining of those objective circumstances. So, by the time he wrote the *Enquiry*, at least, Hume seems to have considered the obtaining of causal relations—and of the causal necessity that is essential to those relations—to be a thoroughly objective matter based on the constant conjunctions and related functional relationships involved (as explored above in key points 1.11 and 1.12 above). Hume's omission of the stridently subjectivist declarations of the *Treatise*, therefore, was both deliberate and appropriate.

2.2. What is the 'Impression' of Necessary Connexion?

Despite all this, however, a suspicion might remain that causal subjectivism was indeed the appropriate conclusion to draw from Hume's empiricist starting point, on the basis that an idea that is copied from an internal impression cannot coherently be ascribed to anything external. So if, on the other hand, Hume wishes to preserve the objectivism of his first definition and rules, then it might seem that he should abandon his empiricist account altogether: what useful role can a subjective impression perform within an objectivist theory?

I shall address this concern by sketching a plausibly Humean account of what his 'impression of power or necessary connexion' might be and how this could generate a corresponding 'idea' that is coherently ascribable to external objects. But to provide independent motivation for this account, it will be useful to start by returning to our earlier discussion and the two highly questionable moves that we saw Hume make when initially framing his impression quest. First (1.6), he casually conflates a wide range of notions, boldly—but somewhat implausibly—claiming 'that the terms of *efficacy, agency, power, force, energy, necessity, connexion,* and *productive quality,* are all nearly synonymous' (*T* 1.3.14.4). Secondly (1.7), he takes for granted that the idea whose impression he seeks is a simple idea despite having previously implied that any such relational idea must be complex (and later going on to assert that any power is relative to its effect). The obvious way of making sense of both of these otherwise gratuitous moves, I suggest, is to interpret Hume as attempting to identify a simple *common element* in all of the various relational notions that he is investigating. When we say that A has an *efficacy, power, force, energy,* or *productive quality* to bring about B, or when we say that A *necessitates* B, we are assigning some kind of *consequential* relation between A and B, a term which is intended to abstract from the detailed differences between these notions and to focus on the fundamental feature that B is understood to be some kind of consequence of A. It is the origin of the idea of this fundamental element of *consequentiality*, I suggest, which is Hume's real quarry.

It is understandable that Hume would view all these consequential notions as problematic from an empiricist point of view, and in exactly the same way: how can any sensory impression or feeling—or even a sequence of such impressions or feelings—possibly give rise to the idea that B was *a consequence of* A, as opposed to merely *following* A? And if this is indeed the fundamental difficulty that motivates Hume's quest, then it also becomes understandable why he might conflate all the various terms and target what he sees as their simple common element. His ingenious innovation is then to switch focus from *causal* consequentiality to *inferential* consequentiality, finding the impression-source of the crucial idea in 'that inference of the understanding, which is the only connexion, that we can have any comprehension of' (*EHU* 8.25). This seems to imply that in customary inference we directly experience a kind of consequential relation within our own minds— awareness of A leads to an expectation of B—this being the only sort of intrinsically consequential 'impression' that our minds ever receive. Though ingenious, however, this answer is itself problematic, because even if our perception of A is regularly followed by our expectation of B, we have no direct awareness of the causal mechanism that underlies this inference, as Hume himself insists (*T* 1.3.14.12; 1.3.14.29; *EHU* 7.9-20). How, then, can this experience of inductive inference help in explaining the impression of causal power?

One possible answer, influentially urged by Barry Stroud (1977, 85–6), is that Hume takes inductive inference to be always accompanied by some distinctive simple feeling, which provides the impression in question. But against this, Hume never says that there is any such 'third perception' between the impression of A and the enlivened idea of B. Such a claim would seem to conflict with what he says about the immediacy and insensibility of inductive inference (*T* 1.3.8.2, 1.3.8.13, 1.3.12.7), and it is hard to see how any such simple feeling—even if it does happen to *accompany* inductive inference—thereby provides an impression of a connexion 'that we can have...comprehension of'. A more attractive resolution, I suggest, is to see Hume as implicitly appealing to a faculty of reflection of a Lockean kind (*Essay* II.i.4, 105–6), which enables us to monitor our mental operations and thus become aware when an inference is taking place rather than simply experiencing a succession of thoughts and feelings. This implies a richer view of reflection than Hume suggests elsewhere in the *Treatise*, where he often writes as though 'impressions of reflection', or 'internal impressions', are confined to 'passions, desires, and emotions' (*T* 1.1.1.1., 7; 1.1.2.1, 11; 1.1.6.1, 16; 1.2.3.3, 27).[27] But clearly some extension of this narrow view is required anyway, if such impressions are to include that of power or necessity.[28] Perhaps this realization explains Hume's change of emphasis in the first *Enquiry*, where what little he says about reflection strongly suggests the Lockean conception of mental monitoring, with talk of 'reflection on the *operations* of our own minds' (*EHU* 7.9, my emphasis) and 'reflection on our own faculties' (*EHU* 7.25) but no mention of the cruder conception—the raw feeling of passions—which had dominated the *Treatise*.[29]

If Hume's account of the impression of power or necessary connexion is indeed informed by this Lockean perspective, then it becomes relatively easy to understand why he so often writes as though the impression is, literally, a 'determination' of the mind or thought, or a customary 'transition of the imagination'.[30] For thus interpreted, the 'impression' is not simply some feeling that happens to accompany inductive inference; rather, it is reflective awareness of such inference taking place, of the very transition itself. This brings at least two considerable advantages. First, it can explain why Hume takes this 'inference of the understanding' to be a 'connexion, that we can have…comprehension of' (*EHU* 8.25), since this form of reflection would enable us to grasp the inference as a movement of the mind from A to B rather than just as a succession of independent perceptions. And that in turn would explain why he sees this as a crucial insight, solving the empiricist conundrum of how consequential concepts can be acquired by experience. Secondly, this account explains how the 'idea' corresponding to that impression might plausibly be seen as essential to a correct understanding of causation (and associated consequential relations) and at least in some sense attributable to external causes and effects such as the motion of billiard balls. If the impression were a mere subjective feeling, then the whole theory would

---

[27] See also *T* 2.3.3.5 and 3.1.1.9, which notoriously suggest a highly atomistic view of these impressions of reflection. Hume's later works make no such atomistic claims and correspondingly downplay his simple/complex distinction, while the related Separability Principle (e.g. *T* 1.1.7.3) disappears.

[28] And also, apparently, that of willing or volition: '*the internal impression we feel and are conscious of, when we knowingly give rise to any new motion of our body, or new perception of our mind*' (*T* 2.3.1.2).

[29] See also: 'the operations of the mind…become the object of reflection' and 'the mind is endowed with several powers and faculties,…[which] may be distinguished by reflection' (*EHU* 1.13–14).

[30] The phrases 'determination of the mind' and 'determination of the thought' occur over a dozen times in the *Treatise*, but never in the *Enquiry*, whereas the phrase 'customary transition' occurs in both works (*T* 1.3.8.11, 1.3.10.9, 1.3.13.3, 1.3.14.24, 1.4.4.1; *EHU* 5.20, 7.28–9, 8.21). I suspect that Hume dropped the term 'determination' in the *Enquiry* because of its causal overtones, which can seem viciously circular when he is trying to account for the origin of our causal concepts, a circularity of which he evinces awareness (*EHU* 8.25 n19).

look bizarre; but if what is being attributed is an inferential relation between events, then it makes far better sense. This in turn renders the theory more plausible interpretatively, because Hume's account of our 'idea of power or necessary connexion' is not intended to debunk that idea. On the contrary, he clearly sees his quest for the crucial impression as successful, and hence as legitimating the corresponding idea through the Copy Principle.[31] It is hard to see how this could be achieved unless the resulting idea is coherently attributable, at least in some sense, to external causes and effects.

2.3. Is Hume a *Projectivist* about Causation (and Morality)?

As remarked above, recent discussions of Hume give considerable prominence to the view that he is best seen as a 'projectivist' about causal necessity, the general idea being that, in ascribing necessity, we are *projecting* onto the external world qualities that are really internal and mental. This, if accepted, might seem to pose a threat to the objectivist theory of causation that I am here attributing to Hume, so I shall briefly explain why I see no such threat.

The idea that Hume is a projectivist is usually combined with the suggestion of a deep parallel between his causal and moral theories, and often motivated by citation of these two famous texts:[32]

'Tis a common observation, that the mind has a great propensity to spread itself on external objects, and to conjoin with them any internal impressions, which they occasion,... the same propensity is the reason, why we suppose necessity and power to lie in the objects,... not in our mind.   (*T* 1.3.14.25)

Thus, the distinct boundaries and offices of *reason* and of *taste* are easily ascertained. The former conveys the knowledge of truth and falsehood: The latter gives the sentiment of beauty and deformity, vice and virtue. The one

---

[31] Hume identifies the impression in question at *T* 1.3.14.20–2 (anticipated at *T* 1.3.14.1) and *EHU* 7.28–30. Hume's attitude to the idea of necessary connexion is thus quite different from his attitude to our thoughts of external objects or selves, which turn out to be fictions rather than bona fide impression-derived ideas (for external objects, see *T* 1.4.2.29, 1.4.2.36, 1.4.2.42–3, 1.4.2.52; for selves, see *T* 1.4.6.6–7). Any interpretation that treats these three topics as together exemplifying a common form of 'Humean scepticism'—or indeed, 'Humean naturalism' (see below p. 232 n37)—should therefore itself be treated with extreme scepticism!

[32] The two passages are cited together by Beebee (2012, 142) and Garrett (2015, 81) and also in the first sentence of Kail's introduction to his book on Humean projection (2007a, xxiii). Blackburn (2008, 27–8) explicates Humean causation as 'a kind of projection of our confidence that one kind of thing will follow another' and goes on to compare this with 'the identical kind of theory that Hume will offer in the case of... ethics.'

discovers objects as they really stand in nature, without addition or diminution: The other has a productive faculty, and gilding or staining all natural objects with the colours, borrowed from internal sentiment, raises, in a manner, a new creation.   (*EPM* App. 1.21)

Since Hume's moral theory has standardly been read as anti-realist, this supposed parallel has encouraged the reading of his causal theory as anti-realist also. But in fact the parallel is questionable, and the pairing of these two famous quotations is highly problematic, because whereas the latter passage apparently approves of 'gilding or staining' in the moral case, the former is clearly critical of mental spreading in the causal case. Indeed, the previous sentence—'This contrary biass is easily accounted for'—makes clear that Hume is here explaining away an erroneous objection to his theory of causal necessity rather than presenting a positive 'projectivist' account.[33]

The second quotation by itself, when seen in context, gives another serious ground for doubt about any would-be projectivist synthesis. For here Hume is distinguishing between the 'boundaries and offices' of reason and taste, saying that reason 'conveys the knowledge of truth and falsehood' and 'discovers objects as they really stand in nature, without addition or diminution' (*EPM* App. 1.21),[34] while taste gilds or stains 'natural objects with the colours, borrowed from internal sentiment'. But, crucially, within this whole discussion Hume himself clearly locates causal judgements within the domain of *reason*,[35] thus standardly representing objects without addition or diminution. By contrast, the sentimental gilding that he associates with taste, and which distinguishes it from reason, appears to involve its action-guiding nature and its association with human desires (*EPM* App. 1.18–20; cf. *T* 3.1.1.6). Thus Hume's distinction between reason and taste here seems to come down to the familiar divide between the cognitive and the conative. And so far from 'gilding or staining' being a unifying theme across Hume's theories of causation and morality, its application to moral judgements is

[33] Likewise, a sentence and footnote elided from the quoted passage make clear that such 'projection' is an error, comparable with attributing spatial location to sounds and smells (Hume alludes here to *T* 1.4.5.11–14).

[34] Reason here is accordingly our cognitive faculty, 'by which we discern Truth and Falshood' (*EHU* 232, 1748 and 1750 editions; cf. *DOP* 5.1), whether of relations of ideas or matters of fact (*T* 3.1.1.9; *EPM* App. 1.6). For extensive discussion of this notion of reason and its relation to 'the imagination' within Hume's thinking, see Millican 2012, 79–85.

[35] *EPM* App. 1.2–3 repeatedly emphasizes that '*reason* instructs us in the several tendencies of actions'. See also: 'the causes and effects...are pointed out to us by reason and experience' (*T* 2.3.3.3), and 'reason, in a strict and philosophical sense,...discovers the connexion of causes and effects' (*T* 3.1.1.12).

precisely what pushes them into the category of taste, thereby distinguishing them sharply from causal judgements, which he clearly takes here to be thoroughly objective and susceptible of truth and falsehood.

3. Hume and Causal Powers

Finally, we come to the question that is particularly germane to the current volume, namely, whether Hume believes that objects have genuine 'causal powers'. I shall argue that he does accept objective powers, though not in the sense that has become prominent in recent decades through the so-called 'New Hume' or *sceptical realist* interpretation. To clear the ground, it will be helpful first to deal with the latter issue.

3.1. Does Hume Believe in 'Thick' Causal Powers?

Sceptical realism—a term coined by John Wright for the title of his 1983 book—involves the claim that Hume believes in a form of objective causal power that is 'thick' in the sense of going beyond his two definitions. On this type of interpretation, which encompasses several varieties, the 'idea of necessary connexion' revealed by his investigations in *T* 1.3.14 and *EHU* 7 does not represent genuine causal necessity at all, but is only a psychological surrogate that manifests our own limited understanding of causation, confined as it is to the observation of regularities and experience of inductive inference. Real causation, by contrast, is usually taken by sceptical realists to involve hidden *absolute* powers or necessities, such that A's being a real cause of B involves A's having some property which, if only we knew of it, would sanction the inference that B must follow *with a priori certainty*.[36] We cannot achieve such perfect knowledge, of course; nor can we form any but the most indirect and relative conception of what such powers and necessities might involve. But—at least on the most prominent of these interpretations—we are still able to believe in them and in the causal relations which they constitute.[37]

---

[36] Strawson (1989, 111) calls this the 'AP property' and Kail (2007b, 256) the 'reference-fixer for power'.

[37] Strawson (1989, 1) particularly emphasizes here what he calls Hume's 'central doctrine of "natural belief"'. But in fact no such doctrine is evident in Hume's texts, and the term derives from Norman Kemp Smith, who saw close parallels between Hume's views on causation and the external world: 'Natural belief takes two forms, as belief in continuing and therefore independent existence,

Most of the evidence adduced in favour of the sceptical realist reading has been problematic and insubstantial, as I have argued at length elsewhere (Millican 2007b; 2009). But it achieved popularity in a context where Hume was widely seen as sceptical about objective causation in general, enabling Galen Strawson's 1989 book *The Secret Connexion* to have a major impact by highlighting a wide range of passages where Hume appears to express sincere objectivist commitments, most notably his references to 'secret powers' in *EHU* 4. Such passages, however, pose no difficulty for the kind of reductionist reading proposed here, encompassing the key interpretative points presented above. For that is itself an objectivist interpretation, explicitly recognizing the search for hidden causes (key points 1.1; 1.5; 1.11) and specifically accounting for Hume's relatively prominent talk of 'powers' in the *Enquiry* (1.12).

There are also a number of powerful objections to the sceptical realist reading, many deriving from its radical conflict with central aspects of Hume's philosophy as generally understood, which provoked Kenneth Winkler's (1991) coining of the moniker 'New Hume' in his eponymous article. To begin with, it requires fundamental reinterpretation of Hume's quest for the impression of necessary connexion, which has to be seen as epistemologically motivated rather than—as the texts themselves suggest (*EHU* 7.3–5, 7.29)—a semantic attempt to define or clarify the meaning of causal terms through identification of the corresponding impression (see Millican 2009, 655–9). On the New view, the genuine causation to which those terms properly refer involves something of which we can have no impression. Yet we do supposedly believe in it, thus apparently violating either Humean empiricism, by allowing ideas that are not derived from impressions, or Hume's theory of belief, by allowing beliefs that are not enlivened ideas.[38] Meanwhile, the type of necessity governing such genuine causation is supposedly absolute and aprioristic, despite Hume's repeated insistence that from an a priori point of view 'Any thing may produce any thing' (*T* 1.3.15.1, 1.4.5.30; cf. *T* 1.3.7.3; Hume *ABST* 11; *EHU* 12.28–9, and see Section 1.4 above). This seems to imply violation of his even more fundamental and oft-repeated Conceivability

---

and as belief in causal dependence' (Smith 1941, 455). For serious doubts about these supposed parallels and the alleged doctrine, see above p. 230 n31 and Millican (2016, 84).

[38] Strawson (1989, 52 and 122) claims that Hume allows a *relative* idea of genuine causation, which could potentially escape this objection because such a complex idea need not be copied directly from any impression. But his account runs into a similar difficulty in explicating the relation involved, as pointed out by Winkler (1991, 62–3) and Millican (2007b, 248 n12). Kail (2007b, 254), in contrast with Strawson, fully recognizes the objection, and accordingly suggests that Hume *assumes* or *supposes* thick powers rather than *believing* in them.

Principle (for discussion see Millican 2009, 676–84), which limits such necessity to matters whose falsehood is inconceivable (*T* 1.3.6.5; Hume *ABST* 11; *EHU* 4.2).

The most serious objections to the New Hume interpretation, however, directly target its central claim that Hume conceives of genuine causation as *thick* in the sense of going beyond his own two definitions. These objections draw attention to various implications that Hume clearly takes to follow from his definitions, all apparently based on the claim that we can have no conception whatever of causal necessity that goes beyond them. Such implications include rejection of 'the common distinction betwixt *moral* and *physical* necessity' (*T* 1.3.14.33) and refutation of the standard argument that 'matter and motion' could not possibly cause thought (*T* 1.4.5.29–33). But the most conspicuous and important application of the two definitions comes in the sections on 'liberty and necessity', with Hume's positive argument that the very same kind of necessity applies to the physical and mental worlds.[39] This crucial argument is essentially the same in the *Treatise*, the *Abstract*, and the *Enquiry*, and in all three it explicitly anticipates protests from those who take physical causation to involve some kind of necessity beyond Hume's two definitions, which is just what the New Hume position implies. Such opponents deny that satisfaction of the definitions 'makes the whole of necessity' (Hume *ABST* 34), 'maintain there is something else in the operations of matter' (*T* 2.3.2.4), and thus 'rashly suppose, that we have some farther idea of necessity and causation in the operations of external objects' (*EHU* 8.22). Hume's response is to insist that his analysis shows any such idea to be 'impossible' (Hume *ABST* 34), and hence that 'there is no idea of any other necessity or connexion in the actions of body' (*EHU* 8.27, cf. *T* 2.3.2.4). He highlights the same point at the beginning of the *Enquiry* version of the argument: 'Beyond the constant conjunction of similar objects, and the consequent inference from one to the other, we have no notion of any necessity, or connexion' (*EHU* 8.5). This passage occurs only six paragraphs after Hume's two definitions of cause (7.29) and is where he starts applying them to solve 'the long-disputed question concerning liberty and necessity' (8.2), 'the most contentious question, of metaphysics, the most contentious science' (8.23). Nobody reading these two sections together could reasonably be in any doubt

---

[39] See Millican 2007a, 190–3; 2007b, 243–5. Beebee (2007) and Kail (2007b, 262–7) proposed answers which were briefly addressed in Millican 2009. Meanwhile Wright (2009, 183–6) suggested that the problem can be circumvented by taking *T* 2.3.2.4 (and presumably similar passages elsewhere) to be disingenuous. The approaches of Beebee, Kail, and Wright were comprehensively criticized in Millican 2011, to which so far no reply has been offered.

that the definitions have been presented expressly with a view to this important application.[40] But in order to serve this role, those definitions have to be understood as delimiting *what we can properly mean* by causal power and necessity, which is exactly what the New Hume interpretation denies. This is as close to an outright refutation as one is likely to find in historical philosophical scholarship.[41]

3.2. Does Hume Believe that Objects Really Have Causal Powers?

If Hume is to be counted as a believer in causal powers, then these must be understood in a way that is compatible with the key points above, strongly suggesting a reductionist approach. Our subsequent discussion has supported this by deflecting the threat of extreme subjectivism, interpreting the 'impression' of causal power or necessity as plausibly attributable to objective occurrences, rejecting any strongly subjectivist form of causal projectivism, and refuting the alternative New Humean understanding of causal powers.

Several of our key points—notably in Sections 1.1, 1.5, and 1.11 above— seem to imply that if we interpret the term 'cause' faithfully to his own theory, Hume does believe in real causes. Since, moreover, he sees causation as essentially involving causal power or necessity (1.3), it seems likewise to follow, again assuming faithful interpretation of the relevant terms, that Hume also believes in real causal powers and real causal necessity.[42] However, there are subtle nuances to be discussed here, as we shall see, and the answer to the question posed is not quite so straightforward.

Identifying the relevant causes, powers, and necessities can be relatively easy where they conform to straightforwardly observable, exceptionless constant conjunctions, and this is the paradigm case from which Hume develops his theory. But as we saw in Sections 1.11 and 1.12, he clearly recognizes, in both the *Treatise* and the *Enquiry*, that it is very far from the whole story.

---

[40] This point has, however, been generally underappreciated, probably owing to most scholars' greater focus on the *Treatise*, where the application to 'liberty and necessity' is postponed until Book Two. The New Humeans' neglect of this application is more surprising, given their emphasis on the *Enquiry* as Hume's authoritative work (Strawson 2000, 31–3; Wright 2000, 95–8).

[41] Other scholars concur: both Ott 2011 and Willis 2015 (205 n43) allude to a general view that 'the New Hume debate has run its course' and been 'ended...once and for all' by the objection from liberty and necessity. Hakkarainen 2012 (307 n36) refers to the objection as 'devastating...against any form of the New Humean interpretation'.

[42] The point about faithful interpretation of the terms is crucial here (see above p. 211 n4). Hume emphatically does not believe in what some other philosophers suppose 'real causal powers' to be.

In more complex cases, the discovery of causal powers and necessities will require careful and painstaking investigation, including systematic observation, experimentation, and generalization, in the attempt to devise laws—sometimes mathematically complex laws—capable of reducing the various phenomena to order. Some of these laws, like the Newtonian laws governing mechanical impact, may involve quantitative factors, such as momentum and kinetic energy, that are naturally expressible in terms of 'energy', 'force', and 'power', through which one can relatively straightforwardly correlate the factor with the effect (thus implying, in the language of *EHU* 7.29 n17, 'that the effect is the measure of the power'). But there is no guarantee that this will always be true,[43] and Hume says so little on these matters that we can only speculate what his reaction would be to yet more sophisticated scientific developments that are not amenable to such expression. Here, however, his apparently crude running together of such a wide variety of causal terms (noted above in Section 1.6) could turn out to be a positive advantage, manifesting his open-mindedness over the form that future theories might take. Rather than attributing to him any firm commitment to 'powers', therefore, it might be better simply to say that he believes the world to have a deterministic causal structure and one that permits—at least to some extent—the human discovery of laws that can predict future outcomes and which hence have a *consequential* nature (in the sense described above).

If the 'power' language of the *Enquiry* is indeed intended to be thus open-minded, then it need not indicate any commitment to *powers in objects*, but only—in a sense—to *objective powers*, powers that are real and not mind-dependent. Accordingly, when Hume talks of the 'powers and forces' of objects, he can be understood as referring to those characteristics—typically unknown and quite likely quantitative—of both individual objects and their situations that determine their behaviour in accordance with the appropriate laws of nature. What then actually occurs will be a holistic result of the entire array of 'powers and forces' operative in the situation, quite different from the simplistic 'inference from one object to another' suggested by his talk of 'constant conjunction', and significantly more complicated than the scenarios envisaged in the *Enquiry* footnotes discussed in Section 1.12, where an object's behaviour is straightforwardly dependent on its own 'power'. But even this, after all, does not represent any wholesale change in approach from

---

[43] Consider, for example, wavelengths of electromagnetic radiation and their interference, in which there will often be no straightforward correlation between quantitative factors and the magnitude of the effect.

Hume's initial presentation of his theory in the *Treatise*, for he was keen from the start to make clear that causal connexion is not to be understood as a property of a single object (*T* 1.3.2.5–6), or even of a cause–effect pair (*T* 1.3.14.15), but rather as involving a relation between kinds of object in the form of a constant conjunction of the cause-kind with the effect-kind (*T* 1.3.14.16), or some more complex relation as envisaged by his rules by which to judge of causes and effects (*T* 1.3.15). Moreover, when presenting his definitions of cause, in both the *Treatise* and the *Enquiry* he draws special attention to this aspect of his theory, noting that identification of a cause can only be made in terms that are 'foreign' to it rather than intrinsic (*T* 1.3.14.31; *EHU* 7.29).[44]

To conclude, therefore, both the early and the mature Hume can whole-heartedly agree that there are objective causal powers in nature, in the sense of stable causal relationships that are mind-independent and subject to truth and falsehood.[45] Some of what he says in the *Enquiry* also seems to endorse attribution of quantifiable powers to individual objects, in cases where we suppose there to be some 'circumstance of an object, by which the degree or quantity of its effect is fixed and determined' (*EHU* 7.29 n17). But the latter will apply only in straightforward cases where 'the effect is the measure of the power', and in more complex scenarios there is unlikely to be any simple correlation between some quantifiable feature of an object and the effect that results from its action. Hence a Humean theory ought to treat attribution of powers to individual objects as dispensable. Fortunately, this accords very well with the spirit of Hume's texts, which, as we have seen, take causal properties to be relational and as arising from patterns of interaction between objects rather than from their individual properties. The upshot is that a consistent Humean—and most likely the historical Hume himself—would be firmly committed to *objective powers*, but not necessarily to *powers in objects*.[46]

---

[44] Perhaps this relational aspect of Hume's theory also made him relatively comfortable with saying that necessity is 'in the mind', since it was then a Lockean commonplace that relations are mind-dependent. Locke (*Essay* II.xxv.8, 322) states that '*Relation* ... [is] not contained in the real existence of Things, and Ephraim Chambers (1738, s.v. 'Relation, *Relatio*')—which takes much of its content from the *Essay*—echoes this: '*relation*, take it as you will, is only the mind; and has nothing to do with the things themselves.' The *Treatise* sometimes seems to follow this orthodoxy, for example by suggesting that a relation 'arises merely from the comparison, which the mind makes' (*T* 1.2.4.21). But more generally, Hume treats relations as thoroughly objective (see Millican 2017, 7–8).

[45] An issue that Hume does not consider is the possibility of multiple theories that are empirically equivalent, so that no one theory is uniquely favoured by the observational data. It seems plausible that, if he had taken this possibility seriously, he would have been inclined towards an instrumentalist approach to the relevant 'powers and forces', preserving the possibility of truth and falsehood even for rival, but equivalent, theories.

[46] For helpful comments on an earlier draft, I am very grateful to Henry Merivale and Hsueh Qu.

References

Beebee, Helen. 2006. *Hume on Causation*. London: Routledge.

Beebee, Helen. 2007. 'The Two Definitions and the Doctrine of Necessity'. *Proceedings of the Aristotelian Society* 107, no. 1.3 (October): 413–31.

Beebee, Helen. 2012. 'Causation and Necessary Connection'. In *The Continuum Companion to Hume*, ed. Alan Bailey and Dan O'Brien, 131–45. London: Continuum.

Beebee, Helen. 2016. 'Hume and the Problem of Causation'. In *The Oxford Handbook of Hume*, ed. Paul Russell, 228–48. Oxford: Oxford University Press.

Blackburn, Simon. 2008. *How to Read Hume*. London: Granta Books.

Chambers, Ephraim. 1738. *Cyclopaedia*. 2nd ed. London.

Clarke, Samuel. 1717. *Remarks upon a Book, Entituled, A Philosophical Enquiry concerning Human Liberty*. London.

Dauer, Francis Watanabe. 2008. 'Hume on the Relation of Cause and Effect'. In *A Companion to Hume*, ed. Elizabeth S. Radcliffe, 89–105 Oxford: Blackwell.

Garrett, Don. 1997. *Cognition and Commitment in Hume's Philosophy*. Oxford: Oxford University Press.

Garrett, Don. 2015. 'Hume's Theory of Causation: Inference, Judgment, and the Causal Sense'. In *The Cambridge Companion to Hume's Treatise*, ed. Donald C. Ainslie and Annemarie Butler, 69–100. Cambridge: Cambridge University Press.

Hakkarainen, Jani. 2012. 'Hume's Scepticism and Realism'. *British Journal for the History of Philosophy* 20 (2): 283–09.

Harris, James A. 2003. 'Hume's Reconciling Project and "The Common Distinction betwixt *Moral* and *Physical* Necessity"'. *British Journal for the History of Philosophy* 11 (3): 451–71.

Harris, James A. 2005. *Of Liberty and Necessity: The Free Will Debate in Eighteenth-Century British Philosophy*. Oxford: Clarendon Press.

Hume, David. 1739–40/2007. *A Treatise of Human Nature: A Critical Edition*. Ed. David Fate Norton and Mary J. Norton. 2 vols. Oxford: Clarendon Press ("*T*").

Hume, David. 1741–77/1985. *Essays, Moral, Political, and Literary*. Ed. Eugene F. Miller. CarmelIndianapolis, IN: Liberty Classics ("*Essays*").

Hume, David. 1748/2000. *An Enquiry concerning Human Understanding: A Critical Edition*. Ed. Tom L. Beauchamp. Oxford: Clarendon Press ("*EHU*").

Hume, David. 1751/1998. *An Enquiry concerning the Principles of Morals: A Critical Edition*. Ed. Tom L. Beauchamp. Oxford: Clarendon Press ("*EPM*").

Hume, David. 1757/2007a. *A Dissertation on the Passions & The Natural History of Religion: A Critical Edition*. Ed. Tom L. Beauchamp. Oxford: Clarendon Press ("*DOP*" and "*NHR*").

Hume, David. 1740/2007b. 'An Abstract of A Treatise of Human Nature'. In A Treatise of Human Nature: A Critical Edition. Ed. David Fate Norton and Mary J. Norton. 2 vols. Oxford: Clarendon Press (as above), 403–17 ("ABST").

Hume, David. 1745/2007c. 'A Letter from a Gentleman to his Friend in Edinburgh'. In A Treatise of Human Nature: A Critical Edition. Ed. David Fate Norton and Mary J. Norton. 2 vols. Oxford: Clarendon Press (as above), 419–31 ("LFG").

Hume, David. 1932. The Letters of David Hume. Ed. J. Y. T. Greig. 2 vols. Oxford: Clarendon Press.

Kail, P. J. E. 2007a. Projection and Realism in Hume's Philosophy. Oxford: Oxford University Press.

Kail, P. J. E. 2007b. 'How to understand Hume's realism'. In Read and Richman 2007, 253–69.

Kames, Henry Home, Lord. 1751/2005. Essays on the Principles of Morality and Natural Religion. 3rd ed. Ed. Mary Catherine Moran. Indianapolis: Liberty Fund.

Locke, John. 1690/1975. An Essay concerning Human Understanding. Ed. P. H. Nidditch. Oxford: Clarendon Press ("Essay").

Mackie, J. L. 1965. 'Causes and Conditions'. American Philosophical Quarterly (2): 245–64.

Mackie, J. L. (1974), The Cement of the Universe: A Study of Causation. Oxford: Clarendon Press.

Millican, Peter. 2007a. 'Humes Old and New: Four Fashionable Falsehoods, and One Unfashionable Truth'. Proceedings of the Aristotelian Society Supplementary Volume 81, no. 1 (June): 163–99.

Millican, Peter. 2007b. 'Against the "New Hume"'. In Read and Richman 2007, 211–52.

Millican, Peter. 2009. 'Hume, Causal Realism, and Causal Science'. Mind 118, no. 471 (July): 647–712.

Millican, Peter. 2010. 'Hume's Determinism'. Canadian Journal of Philosophy 40, no. 4 (December): 611–42.

Millican, Peter. 2011. 'Hume, Causal Realism, and Free Will'. In Causation and Modern Philosophy. Ed. Keith Allen and Tom Stoneham, 123–65. New York: Routledge.

Millican, Peter. 2012. 'Hume's "Scepticism" about Induction'. In The Continuum Companion to Hume. Ed. Alan Bailey and Dan O'Brien, 57–103. London: Continuum.

Millican, Peter. 2014. 'Skepticism about Garrett's Hume: Faculties, Concepts, and Imposed Coherence'. Hume Studies 40, no. 2 (November): 206–26.

Millican, Peter. 2016. 'Hume's Chief Argument'. In The Oxford Handbook of Hume. Ed. Paul Russell, 82–108. Oxford: Oxford University Press.

Millican, Peter. 2017. 'Hume's Fork, and His Theory of Relations'. *Philosophy and Phenomenological Research* 95, no. 1 (July): 3–65.

Ott, Walter. 2011. Review of *Causation and Modern Philosophy*, ed. Keith Allen and Tom Stoneham. *Notre Dame Philosophical Reviews*, 4 July 2011. http://ndpr.nd.edu/news/causation-and-modern-philosophy/.

Read, Rupert, and Kenneth A. Richman, eds. 2007. *The New Hume Debate*. Rev. ed. London: Routledge.

Smith, Norman Kemp. 1941. *The Philosophy of David Hume*. London: Macmillan.

Strawson, Galen. 1989. *The Secret Connexion: Causation, Realism, and David Hume*. Oxford: Clarendon Press.

Strawson, Galen. 2000. 'David Hume: Objects and Power'. In Read and Richman 2007, 31–51.

Stroud, Barry. 1977. *Hume*. London: Routledge & Kegan Paul.

Willis, Andre. 2015. *Toward a Humean True Religion*. University Park: Pennsylvania State University Press.

Winkler, Kenneth P. 1991. 'The New Hume'. In Read and Richman 2007, 52–74.

Wright, John P. 1983. *The Sceptical Realism of David Hume*. Manchester: Manchester University Press.

Wright, John P. 2000. 'Hume's Causal Realism: Recovering a Traditional Interpretation'. In Read and Richman 2007, 88–99.

Wright, John P. 2009. *Hume's A Treatise of Human Nature: An Introduction*. Cambridge: Cambridge University Press.

10

Resurgent Powers and the Failure
of Conceptual Analysis

Jennifer McKitrick

The ontological status of potentialities, powers, and dispositions has been debated in philosophy for centuries. On some accounts of this history, powers came to be regarded as unfit for serious ontology and relegated to the dustbin of philosophical history. But in the twenty-first century, we find a profusion of philosophical literature on powers, raising the question: if powers were in the dustbin of history, how did they get out? Part of the answer is that philosophers, scientists, and laypeople never stopped making disposition ascriptions: they continued to describe things as malleable, conductive, soluble, volatile, etc. If we regularly find our linguistic community making assertions about the dispositions of things, what should that mean for our ontology? Three alternatives come to mind. We could take these assertions at face value and take them to commit us to the existence of dispositions. Or, we could adopt an error theory, concluding that all such assertions are false. Alternatively, we could try to explain how some of these assertions can be true, consistent with the non-existence of dispositions. One way that disposition ascriptions can be true even if dispositional properties do not exist is if those assertions can be shown to be semantically equivalent to assertions that are devoid of dispositional predicates. However, the project of attempting to eliminate dispositional predicates from ordinary, philosophical, and scientific discourse has a troubled past.

This chapter will retrace the twentieth-century research programme of analysing disposition ascriptions (Malzkorn 2001; Schrenk 2009; Choi and Fara 2012), first in terms of material conditionals, then in terms of counterfactual conditionals. Many philosophers deem this a failed project. Since it seems unlikely that disposition ascriptions can be eliminated from our language, many have concluded that dispositional predicates refer to real

Jennifer McKitrick, *Resurgent Powers and the Failure of Conceptual Analysis* In: *Reconsidering Causal Powers: Historical and Conceptual Perspectives*. Edited by: Benjamin Hill, Henrik Lagerlund, and Stathis Psillos, Oxford University Press (2021).
© Jennifer McKitrick.
DOI: 10.1093/oso/9780198869528.003.0011

dispositional properties, powers, or powerful properties—properties that have causal 'oomph'. According to these neo-Aristotelian approaches, the world is teeming with powerful properties. Powers have come roaring back! But perhaps there has been an overreaction to the apparent failure of conditional analyses. In the end, I recommend a more moderate and epistemically modest stance toward causal powers.

1. Reducing Dispositions in the Twentieth Century: a Guide for Beginners

1.1. Logical Positivism

Why are so many twentieth-century philosophers suspicious of dispositions? One reason harkens back to Hume's view (T 1.3, 50–120) from an earlier century that we can have no empirical evidence for causal powers in nature, nor necessary connections between distinct existences. If dispositions are supposed to be causal powers necessarily connected to their manifestations, then, from a certain empiricist point of view, we have no justification for positing their existence.

A related source of suspicion is the epistemological outlook most closely associated with logical positivism. According to the logical positivist's principle of verification (Ayer 1936), every meaningful statement must be verifiable, either empirically or analytically. Disposition ascriptions, such as 'that tablet is water-soluble', are clearly not analytic, and so they must be empirically verifiable. So, how does one empirically verify that a tablet is water-soluble? Unlike 'white' or 'round', 'water-soluble' is not an observation term—you can't tell that something is water-soluble merely by examining its current state. Furthermore, a disposition that is not manifesting seems to be related to a merely possible state of affairs. For example, one might think that the water-solubility of the tablet stands in some relation to dissolving, but if it is undissolved, that dissolving is a merely possible event or uninstantiated property. If relations only hold between things that actually exist, the idea of properties that reach beyond themselves to merely possible events or property instances seems problematic (Armstrong et al. 1996, 17).

But disposition terms, such as 'soluble', 'electrically conductive', 'malleable', etc., are common in scientific discourse. Given the logical positivists' veneration of scientific methods, dismissing dispositional vocabulary as meaningless is not an option. So, ascribing a dispositional predicate to a substance must be a way of saying something empirically respectable. That is, disposition ascriptions must be translatable or semantically reducible to other statements employing nothing but observation terms. Similar to the way logical positivists treat other theoretical terms such as 'temperature', disposition terms are rendered acceptable via reduction sentences or operational definitions. If a term is operationally defined, then ascribing that term to an object is to assert that, if a certain test or operation is performed on that object, then there will be a certain observable result. (The operation that is performed on the object is also known as the disposition's 'test', 'stimulus condition', 'trigger', or 'circumstance of manifestation'.) So, to say that a tablet is water-soluble is to say something like 'if the tablet is submerged in water, then it will dissolve'. If 'tablet', 'dissolve', etc. are not observation terms, then further reducibility is required.

Operational definitions are perhaps the earliest attempts to reduce disposition ascriptions to conditionals. A straightforward articulation of this idea, contemplated by Rudolph Carnap (1936, 440), is the Naive Conditional Analysis: (NCA) x has disposition $D =_{df}$ For all times t (If operation O is performed on x at t, then x gives response R at t). Given the state of logic at the time, the 'if-then' statement is taken to be a truth-functional, material conditional. Material conditionals have the noteworthy feature of being true whenever the antecedent and consequent are both true, whether or not there is any causal connection between them. This feature leads to the problem of random coincidences, where 'operation O is performed on x' and 'x gives response R' just happen to both be true at and only at time t. For example, suppose 'x is fragile' is defined as 'if x is struck at t, then x shatters at t'. Now suppose that an iron safe has a time bomb in it, and the only time the safe is struck coincides with the moment that the bomb explodes. So, for all times t, the safe is struck at t, and it shatters at t. It follows from NCA that the iron safe was fragile, and that is an undesirable result. Another noteworthy feature of the material conditional is that the whole conditional is true when the antecedent is false. This feature creates another problem for NCA—void satisfaction. NCA entails that all objects that never have a certain operation performed on them have the associated disposition. For example, everything that is not submerged in water turns out to be water-soluble. As a result, Carnap (440) concludes that dispositional predicates cannot be defined by NCA 'nor by any other definition'.

Carnap (443) then attempts to provide a stipulative definition of a disposition term by means of a Bilateral Reduction Sentence: (BRS) For all x and all times t [if O is performed on x at t then (x has D if and only if x gives R at t)].

However, BRS also has problems. First, BRS does not avoid the problem of random coincidences. Consider the case of the iron safe described above. If t is the only time that 'striking' is performed on the iron safe, then the safe is fragile if and only if it shatters at t. Since the safe does shatter at t, BRS counts the iron safe as fragile. The second problem is that it does not provide a way of determining the truth-value of disposition ascriptions in untested cases. If 'O is performed on x at t' is false, then the whole conditional is true, whether or not 'x has D if and only if x gives R at t' is true. So, if O is never performed on x, there is no way to determine whether x has D. Third, if the object fails a single test, that is, if x does not give R when O is performed on x, it turns out that x does not have the disposition. So, if you strike a match once and it does not catch on fire, it would follow that the match is not flammable—another unwanted result.

Other reductionists attempt to devise better definitions of disposition terms, and in doing so they introduce further complexities, such as appealing to an object's similarity to other objects that exhibit the response in question. Consider, for example, the Kaila/Storer (Kaila 1939; Kaila 1941; Storer 1951; discussed by Malzkorn 2001) definition of solubility: (KSS) x is soluble $=_{df}$ x is in water and x dissolves, or there exists some predicate 'is F' such that [x is F and there is some y such that (y is F and y is in water and y dissolves) and there is no z such that (z is F and z is in water and z does not dissolve)]. But despite its complexity, KSS suffers the same problem as NCA—void satisfaction. Since there are no restrictions on which predicates that can serve as the value of 'is F', we can use a predicate such as 'is identical to either x or y', which only applies to x and y. Then, as long there is some y that has dissolved, any x that is never submerged counts as soluble (Malzkorn 2001, 340).

1.2. Gilbert Ryle

The story of the philosophical discussion of dispositions in the twentieth century would not be complete without mentioning the work of Gilbert Ryle (1949, discussed in Scholz 2009), not because Ryle offers a new way to analyse disposition ascriptions, but because he articulates another motivation for doing so, a motivation that proves to be very influential for decades to come. Ryle's characterization of dispositions is central to his attack on mind–body dualism. Ryle argues that attributions of psychological characteristics to people are tantamount to disposition ascriptions. Given Ryle's desire to repudiate any type of inner, private mental entities, such a disposition ascription's being applicable to a person is not meant to entail that the person instantiates some dispositional *property*. Rather, to attribute a disposition to someone is merely to assert that a certain conditional is true of them, basically to assert that you predict that they will behave in certain ways in certain circumstances. A disposition ascription is thought to be merely an 'inference ticket' that gives you permission to infer a conclusion about a thing's behaviour from your premises about its circumstances (Ryle 1949, 121).

The idea that mentality is importantly similar to dispositionality, and that philosophical progress on both of these topics can be made by attending to this similarity, is one that would persist (Armstrong 1973; Block 1990; Jackson 1995; Shoemaker 2001). However, Ryle's association of mental concepts with dispositions does nothing to solve the outstanding difficulties for analysing disposition ascriptions. To these problems, Ryle adds the complication of 'multitrack' dispositions. Where a 'single-track' disposition has one type of stimulus and one type of manifestation, a multitrack disposition can by stimulated by different kinds of circumstances and manifest in different ways. Ryle acknowledges the problems that multitrack dispositions create for the prospects of semantic reduction when he writes:

> There are many dispositions the actualizations of which can take a wide and perhaps unlimited variety of shapes.... If we wish to unpack all that is conveyed in describing an animal as gregarious, we should similarly have to produce an infinite series of different hypothetical propositions.
>
> (Ryle 1949, 43–4)

1.3. Nelson Goodman

By the 1950s, philosophers recognize that material conditionals are inadequate for analysing disposition ascriptions. So, they introduce stronger-than-material conditionals, such as causal implication or counterfactual conditionals backed up by natural kinds and laws of nature. But such posits threaten to abandon some basic tenets of empiricism (Pap 1958).

Goodman (1979) offers a glimpse into the philosophical discussion of dispositions in the 1950's. Goodman considers the problem of analysing dispositions to be among the most urgent and pervasive in epistemology and philosophy of science (33). Goodman famously echoes the reductionist aspirations of the logical positivists: 'The dispositions or capacities of a thing—its flexibility, its inflammability, its solubility—are no less important to us than its overt behavior, but they strike us by comparison as rather ethereal. And so we are moved to inquire whether we can bring them down to earth; whether, that is, we can explain disposition-terms without any reference to occult powers' (40). However, Goodman thinks the problem is even more pervasive than that, since he thinks that 'almost every predicate commonly thought of as describing a lasting objective characteristic of a thing is as much a dispositional predicate as any other' (41).

Goodman is highly critical of contemporaneous accounts of dispositional predicates. He characterizes the view that disposition terms are undefinable primitives as 'so prevalent' that it 'must be dealt with' (46). The support for this Carnapian view, according to Goodman, is merely the assertion that it accurately represents scientific practice. Goodman rejects this justification, as it is the function of philosophy to explicate scientific language, not merely to depict it (47). He rejects the proposal that a dispositional ascription is 'a summary description of certain aspects of the total history of a thing' because 'the defects in this too-simple proposal are well-known': Like NCA, this proposal fails when it comes to dispositions that are never tested (42–3). Another view that Goodman considers is the following: 'If among things under suitable pressure, "flexes" applies to all and only those that are of kind K, then "flexible" applies to all and only those that are of kind K whether they are under pressure or not' (44). Goodman finds this view to be wanting due to the lack of an account of natural kinds. Defining natural kinds in terms of essential properties would not help, because there is no reason to assume that a thing's dispositions are essential to it. For example, a portion of frozen water can be disposed to shatter when struck, but that disposition is not essential to it, since the water can melt. Perhaps one could identify the kind *water* by its microstructure—being H_2O. But identifying kinds by their microstructures will not help the reductive project, Goodman (45) argues, since the relevant predicates that describe microstructures are dispositional as well.

Goodman's own tentative suggestion about how to analyse dispositions is the following: 'We can define "flexible" if we find an auxiliary manifest predicate that is suitably related to "flexes" through "causal" principles or *laws*' (45). To elaborate, in contrast to dispositional predicates, to apply a 'manifest predicate' to something 'is to say that something specific actually happens to the thing in question' (41). 'Breaks', 'burns', 'dissolves', and 'looks orange' are given as examples of manifest predicates. So, suppose some manifest predicate 'is P' is applicable to a certain object, and 'is P' is causally or nomically connected to another manifest predicate, 'flexes'. Then, according to Goodman's proposal, the predicate 'flexible' is applicable to that object. As

Goodman acknowledges, this would give us, at best, a definition of 'flexible', and not a general formula for defining dispositional predicates. Defining other dispositional predicates will have to proceed in a piecemeal fashion. But more problematically, Goodman notes 'the question of when such a "causal" connection obtains or how laws are to be distinguished from accidental truths is an especially perplexing one' (46–7). So, by Goodman's *own* lights, his own proposal does not fare much better than its rivals.

The proposal for analysing disposition ascriptions that receives the most attention in *Fact, Fiction, and Forecast* involves counterfactual conditionals. Goodman reports that it is 'a common habit of speech' and 'a recent trend in philosophy' to express ideas about dispositions 'in counterfactual form' (34). This suggests replacing 'k was flexible at time t' with 'if k had been under suitable pressure at time t, then k would have bent' (34–5). Goodman cites the virtues of this replacement: 'The disposition term "flexible" is eliminated without the introduction of any such troublesome word as "possible"; only non-dispositional predicates remain' (35). However, Goodman abandons the 'familiar and inevitable' (43) counterfactual analysis since 'we are still a very long way from having a solution to the problem of counterfactuals…after a number of years of beating our heads against the same wall and of chasing eagerly up the same blind alleys'. Consequently, 'no ground is gained by taking the problem of counterfactuals in trade for the problem of dispositions' (38).

To get a better sense of this problem, we must look back to the first essay of *Fact, Fiction, and Forecast,* entitled 'The Problem of Counterfactual Conditionals'. There Goodman claims that, without an adequate analysis of counterfactual conditionals, 'we can hardly claim to have any adequate philosophy of science' (38), which must include an account of disposition terms. Goodman characterizes a challenge for interpreting counterfactuals as follows: 'Plainly, the truth-value of the counterfactual does not derive simply from the truth-value of its components; for since the antecedent and the consequent of every counterfactual are both false, all counterfactuals will have the same truth-value by any truth-functional criterion' (36). Goodman uses the term 'counterfactual' to refer to claims that are literally contrary to the facts, but even if we construe it more broadly, the point remains: the truth-value of a counterfactual is not a function of the truth-values of its antecedent and consequent. There is clearly an important relation between the antecedent and the consequent of a counterfactual, but the nature of this relation is unclear. The antecedent of a counterfactual does not logically entail its consequent. Goodman explores the idea that the antecedent *plus relevant conditions and laws* jointly and non-trivially entail the consequent. However, identifying the relevant conditions proves to be problematic. Consider: if the match had been scratched, it would have lit. What are the relevant conditions that need to be added to the antecedent to entail the consequent? Goodman considers adding all true statements, for surely that would be more than enough. However, the antecedent plus *all* true statements will not do, because all true statements include the statement 'the match was not struck'. So, negations of the antecedent must be excluded from the relevant conditions, along with anything that logically entails the negation of the antecedent, and anything that is physically incompatible with the antecedent (9–15). In short, the relevant circumstances must be 'cotenable' with the antecedent of the counterfactual. However, Goodman laments:

> In order to determine the truth of a given counterfactual it seems that we have to determine, among other things, whether there is a suitable [set of true sentences] S that is cotenable with [the antecedent] A.... But in order to determine whether or not a given S is cotenable with A, we have to determine whether or not the counterfactual 'If A were true, then S would not be true' is itself true...we can never explain a counterfactual except in terms of others, so that the problem of counterfactuals must remain unsolved.   (16–17)

Furthermore, even if the relevant conditions could be identified, analysing counterfactuals in terms of conditions and laws still requires an account of scientific laws, the development of which is stymied by Goodman's infamous worries about induction (17–27; 59–83).

In short, Goodman evaluates available philosophical accounts of dispositions and finds them all to be lacking. Proposed definitions of dispositions appeal to laws, causal connections, natural kinds, or essences, and each of these concepts is problematic in its own right, from an empiricist point of view. Nevertheless, this reading of *Fact, Fiction and Forecast* suggests that the main obstacle to reducing dispositional predicates is considered to be the lack of an account of counterfactual conditionals. Once philosophers have an account of counterfactuals, a counterfactual analysis of dispositions is largely assumed without argument.

1.4. Possible Worlds Semantics

Important work on counterfactuals by Robert Stalnaker (1968) and David Lewis (1973) enters the discourse in the early 1970s. The tool that systematizes the interpretation of counterfactual conditionals is a multiverse of possible worlds. According to this approach, there are as many possible worlds as there are ways the world could be or might have been. Possible worlds can be compared with respect to overall similarity or 'closeness'. The basic idea for interpreting counterfactuals is a counterfactual is true if and only if, in the closest possible world in which the antecedent is true, the consequent is true.[1] While the term 'counterfactual' does imply that the antecedent is contrary to the actual facts, that implication is not assumed in this account. The closest possible world in which the antecedent is true might be the actual world. With the availability of possible worlds semantics for counterfactuals, a major barrier to reductively analysing dispositions ascriptions is lifted. If disposition ascriptions are equivalent to counterfactuals, that means that, in the closest possible world where the disposed object is in certain circumstances, a certain manifestation occurs. For example, to say that a glass is fragile is to say that in the closest possible world in which the glass is struck, the glass breaks.

While possible worlds semantics offers the promise of reducing disposition ascriptions, we need to pause here a moment and note how far this reductionist project has come from its Humean roots. Depending on the sense that can be made of possible worlds talk, it seems like the remedy is worse than the disease, from an empiricist point of view. If dispositions are problematic because they are 'too ethereal', to use Goodman's phrase, then possible worlds should be anathema. We started off questioning the empirical respectability of a salt tablet's water-solubility and wound up with infinitely many possible worlds containing infinitely many objects with which we have no causal interaction, and for which we have no empirical evidence. While Stalnaker and others (Stalnaker 1976; Armstrong 1989a; Rosen 1990) try to find an interpretation of possible worlds talk that is not ontologically extravagant, Lewis (1986) embraces the existence of infinitely many possible worlds that do not differ in kind from the actual world; the actual world is distinguished only by the fact that it is *our* world.

Despite this metaphysical extravagance, Lewis associates himself with the empiricism of the past, entitling one of his main theses 'Humean supervenience'—the view that, in worlds like ours, everything supervenes on local matters of particular fact. Accordingly, anything that is true in the actual world, including causal claims, laws of nature, and disposition ascriptions are true in

---

[1] I am simplifying. On Lewis's (1973, 16) view, the sentence 'If it were the case that *P*, then it would have been the case that *Q*' would get analysed as 'Either there is no close enough world where it is the case that *P*, or there are close enough worlds where it is the case that *P*, and in all of the closest possible worlds where *P* is true, *Q* is true.'

virtue of the intrinsic qualities of fundamental, possibly point-sized entities: there are no 'necessary connections' (1983–6, 2:ix). We will get to Lewis's analysis of disposition ascriptions shortly, but notice Lewis's Humean super-venience is a metaphysical claim, not an epistemic one. Lewis does not merely claim that empirical evidence does not support the existence of necessary connections in nature but makes the positive proposal that there are none in the actual world. This move, from the (purported) absence of evidence to the evidence of absence, provides yet another motivation for reducing disposition ascriptions: there can't be real dispositional properties because there *are* no necessary connections in nature. Consequently, true disposition ascriptions must be semantically equivalent to claims without any apparent reference to dispositional properties, and counterfactuals which mention only circum-stances and manifestations appear to be suitable for that role.

1.5. The Simple Counterfactual Analysis

If the major impediment to reductively analysing disposition ascriptions previous to 1970 is the lack of an account of counterfactuals, recent devel-opments suggest that a reductive analysis should now be relatively straight-forward. A Simple Counterfactual Analysis (SCA) equates a disposition ascription with a simple counterfactual conditional as follows: (SCA) x has D $=_{df}$ If x were in C, x would exhibit M. For example: x is fragile $=_{df}$ If x were struck at t, x would shatter at t. Now there is no problem of void satis-faction. It is not true of all things that aren't struck that, if they were struck, they would break. Consider a wooden log that is not being struck. The closest possible world in which the log is struck is otherwise like the actual world. So, it is reasonable to suppose that the wooden log does not break in that world, despite being struck. Therefore, the log is not fragile in the actual world.

However, there is still a problem of random coincidence. Recall the safe that is struck at the moment a time bomb inside of it explodes. The counter-factual 'if it were struck at *t*, it would shatter at *t*' is true of the safe since, in the closest possible world in which the antecedent is true (that is, the actual world), the consequent is also true. According to Lewis's (1973, 3) analysis, a counterfactual that has a true antecedent and a true consequent is true. So, according to SCA, the iron safe was fragile. One can avoid this result if one claims that there are other possible worlds that are just as similar to the actual world as the actual world is to itself (Lewis 1973). But that seems ad hoc and implausible. And that is not the only problem with SCA, as many counterexamples are brought to bear against it.

2. Counterexamples to the Simple Counterfactual Analysis

Around the turn of the twenty-first century, philosophical discussions about dispositions feature a variety of counterexamples to SCA, some fairly straightforward, others clever and fantastical. While any particular purported counterexample might not be decisive on its own, the ongoing barrage proves to be a continuous challenge. Despite their apparent variety, these cases are now generally recognized to be of three basic types: mimics, masks, and finks.

2.1. Mimics

Mimics are objects that will exhibit a manifestation distinctive of a certain kind of disposition in the relevant circumstances even though they do not possess that disposition. So, mimics show that a counterfactual holding of an object is not sufficient for it having the associated disposition. A. D. Smith (1977) offers the example of a wooden block that would break when dropped on Neptune, due to something peculiar about the planet's atmosphere. If the block is brought to Neptune, the counterfactual 'if it were dropped, it would break' would be true, even though the wooden block is not fragile. Another mimic case involves the 'hater of styrofoam', a person who tears up nearby styrofoam cups whenever they are struck (an example attributed to Daniel Nolan in Lewis 1997). The closest possible world in which the styrofoam cup is struck need not be the actual world, but it will be one in which the hater of styrofoam is nearby. Consequently, the cup will be broken in that world, and so it will count as fragile in the actual world. Mimic cases (called 'regulated coincidences' by Schrenk 2009, 151) are similar to random coincidences, but the correlation is not totally random, but regulated. The counterfactual that is true of the mimic can have a false antecedent. Consequently, even if there is a better way to deal with counterfactuals with true antecedents (McGlynn 2012), that will not address the problem of mimics.

A defender of SCA might say that a complete specification of the circumstances of manifestation for a given disposition will rule out extrinsic factors like the hater of styrofoam. But that will not work for a kind of mimic case where the mimicking factor is intrinsic to the mimic. Manley and Wasserman

(2008, 67) offer such a case that they call 'Achilles' heel'. A wooden block is relatively sturdy, but it has one weak spot. If it is dropped and lands on that spot, it will break. Since the weak spot is intrinsic to the block, getting more specific about the external circumstances will not help.

2.2. Masks and Antidotes

Other counterexamples show that the truth of the counterfactual is not necessary for the truth of the disposition ascription. Even when a disposed object is in the circumstances of manifestation, some interference can prevent the disposition from manifesting. These are known as masks or antidotes. A fragile thing can be protected with bubble wrap so that, if it were struck, it would not break. And 'cyanide is lethally poisonous to humans' is not equivalent to 'if a human ingested cyanide, he would die', because the consequent would be false in the case where the human also ingested an antidote to cyanide. Another example, provided by Bird (1998), is the plutonium in a nuclear reactor that is disposed to explode. The counterfactual 'if the plutonium started a nuclear reaction, then it would explode' is not true since the nuclear reactor is built with boron rods and other fail-safes, which slow down the reactions, preventing explosions. Manley and Wasserman (2008, 69) offer a twist on masking with their example of a glass that has a 'reverse Achilles' heel': 'it can withstand a surprisingly strong force, provided that the force is applied at *exactly* the right angle and at *exactly* the right point. Despite the reverse Achilles' heel, the glass is extremely fragile.' The glass's reverse Achilles' heel masks its fragility when it is hit in that specific way. As above, the fact that the reverse Achilles' heel is intrinsic to the glass causes problems for the obvious strategies for ruling out masks, such as getting more specific about the circumstances of manifestation.

2.3. Finks and Alterations

Another kind of counterexample to SCA involves altering the disposed object when it is in the circumstances of manifestation in such a way that the object loses its disposition, and consequently the manifestation does not occur. The dispositions instanced in these kinds of cases are known as 'finkish dispositions' in the literature, after C. B. Martin's (1994) example which he calls 'the electro-fink'. Martin's electro-fink is a machine that can turn a live electrical wire into a dead one and vice versa. Martin considers the following conditional analysis of 'the wire is live': 'if the wire is touched by a conductor then electrical current flows from the wire to the conductor' (2). However, if the wire is connected to the electro-fink, then if the wire is touched by the conductor, the electro-fink will make the live wire dead. While the example may seem cooked up, as David Lewis (1997, 147) notes, the electro-fink 'need not be anything more remarkable than a (sensitive and fast-acting) circuit-breaker'.

Putting Martin's idea in other words, an object has a finkish disposition if that object has a disposition which it would lose if the circumstances of manifestation for that type of disposition were to occur. Recall SCA: [A] x has D $=_{df}$ [B] If x were in C, x would exhibit M. If disposition D is finkish, the same circumstance C that would typically cause objects with D to exhibit M instead causes x to lose D before it can exhibit M. In this case, [A] is true: x does have the disposition. But [B] is false: If x were in C, it would not exhibit M, since x would lose D first. So, SCA fails to give a necessary condition for having that disposition. Finks differ from masks because subjecting an object with a masked disposition to the circumstances of manifestation does not alter the disposition, but merely prevents it from manifesting. Subjecting an object with a finkish disposition to the circumstances of manifestation destroys the disposition.

Other examples of finkish dispositions include the fragility of a glass which is protected by a sorcerer who will immediately render it non-fragile if it is ever struck (Martin 1994, 2; Lewis 1997). For a case where the loss of the disposition is associated with easily observable changes, Martin (1994, 4–5) offers the 'moleculo-fink'. The moleculo-fink is hooked up to a fragile block of ice and will instantly melt the ice if it is ever struck. Mark Johnston (1992) gives us the example of the shy, but intuitive, chameleon. Assuming a dispositional account of colour, the green chameleon is disposed to look green, but before anyone can turn on the light and look at it, it blushes red. Yet another example is 'killer yellow'. Normal perceivers are disposed to see yellow when they look at yellow things, but should a person ever look at the killer yellow object, she would be killed instantly, before she had any visual experience of yellow (attributed to Saul Kripke by Lewis 1997, 145). A less fantastical example of a finkish disposition (Tornaletti and Pfeifer 1996) is the instability of the DNA molecule. DNA is susceptible to breaking up due to forces such as radiation and heat. However, forces which would break the molecule also trigger mechanisms within the cell nucleus that help maintain the molecule's structure.

A thing can also finkishly lack a disposition or be subject to a 'reverse fink'. A dead wire that is hooked up to the electro-fink finkishly lacks the disposition of being live. When the stimulus condition for being live (being touched by a conductor) occurs, the electro-fink causes the dead wire to become live. When green, the shy but intuitive chameleon finkishly lacks a red appearance. When the circumstances of manifestation are about to occur (someone is going to look at him), the chameleon acquires a red appearance. If a dispositional account of colour is correct, the chameleon acquires a new disposition—a first-order disposition to appear red. Reverse finks differ from mimics because an object that mimics having a disposition does not acquire it, whereas the reverse fink makes it the case that an object without a certain disposition acquires that disposition in the circumstances of manifestation. In these cases, an object x which does not have disposition D gains D when exposed to circumstance C, and subsequently exhibits manifestation M. Arguably, [A] is false: x does not have the disposition. However, [B] is true: if x were to be subject to C, x would exhibit M. This shows that SCA fails to give a sufficient condition for x's having a disposition. Such counterexamples prompted David Lewis (1997, 143) to declare: 'The simple conditional analysis has been decisively refuted by C. B. Martin.'

3. Sophisticated Counterfactual Analyses

SCA has some defenders (Gundersen 2002; Choi 2006; Choi 2008), but the more common response is to say that the simple counterfactual analysis is just too simple. Some philosophers hope that a more complex analysis is not falsified by mimics, masks, and finks. Numerous sophisticated analyses have been put forward, with various costs and benefits.[2] However, due to considerations of space, I will assess only a sample of the most influential approaches.

3.1. *Ceteris Paribus*

Goodman (1979, 39) notes in passing that it is possible that a dry piece of wood w is inflammable, while the counterfactual 'if w had been heated enough, it would have burned' is false because w is in an atmosphere with no

---

[2] These include appeal to ideal conditions, conditional conditions (Mumford 1998), second-order properties, and functional properties (Prior 1985).

oxygen. Goodman claims that this shows that the disposition ascription is, at best, translatable into some 'fainthearted counterfactual' including a clause such as 'if all conditions had been propitious' (39). This suggests that the counterfactuals that might plausibly define dispositional predicates must contain *ceteris paribus* clauses (Prior 1985; Lewis 1997; Bird 1998; Mumford 1998; Malzkorn 2000; Mellor 2000; Cross 2005; Choi 2006; Choi 2008; Hauska 2008; Steinberg 2010). Perhaps such clauses will not only ensure that there is oxygen around, but also that there is no masking, mimicking, or finking going on. Call the amended account the Ceteris Paribus Counterfactual Analysis: (CPCA) x has D $=_{df}$ If x were in C then, if other things were equal, x would exhibit M. Masking, such as wrapping a glass in bubble wrap, is not keeping other things equal, so a fragile glass's failure to break when struck when wrapped in bubble wrap is no counterexample. Similarly, mimicking, such as taking a wooden block to Neptune, or having a hater of styrofoam around, would not be keeping other things equal. However, even if CPCA effectively deals with masks and mimics, the problem of finks may prove to be recalcitrant.

Martin (1994, 5) considers an account like CPCA when he writes, 'The conditional *analysans* considered above is too simple. Nobody believes, or ought to believe, that the manifestation of powers follow upon a single event mentioned in the antecedent of the conditional, independently of what the circumstances are. Conditionals which give the sense of power ascriptions are always understood to carry a saving clause (the full details of which are commonly not known).' Keeping with Martin's example, a proponent of CPCA would say: [A] The wire is live is equivalent to [B']: If the wire is touched by a conductor and other things are equal, then electrical current flows from the wire to the conductor. Having the wire hooked up to the electro-fink is not keeping other things equal, so Martin's case is not supposed to be a problem.

However, according to Martin this does not work. It's not the case that you have to keep *everything* the same, since irrelevant differences should not falsify the antecedent of the counterfactual. What must be kept the same is, at a minimum, that the wire is not hooked up to the electro-fink. But also, we need to rule out anything that accomplishes the same thing that the electro-fink does. That is, in order to keep 'other things equal', we must exclude anything that would do what a fink does. And we could never be sure if we had an exhaustive list of such things, or even a characterization of their general features. So, the only way to characterize all of those things is in terms of what they do. And what do all finks do? For one thing, they falsify the counterfactual. If there is a fink about, then some proposition that is inconsistent with the counterfactual is true. So perhaps CPCA effectively comes to the following: (CPCA′) x has D =$_{df}$ If x were in C, x would exhibit M, unless some proposition P is true, where [if P, then it is not the case that (if x were in C, x would exhibit M)]. The problem with this understanding of the CP clause is that it trivializes the analysis to the point where everything has disposition D, since everything is such that the counterfactual is either true of it or not. To avoid this result, we might turn to the other thing that a fink does—it removes the disposition. So, to rule out finks, one thing that must be kept equal is that the object keeps the disposition. The resulting analysis is the following: (CPCA″) x has D =$_{df}$ If x were in C, x would exhibit M, unless some proposition P is true, where [if P, then (if x were in C, then x would not have D)]. If this is the most plausible interpretation of CPCA, since it mentions the disposition D in the analysans, then it is not a reductive analysis capable of explaining away dispositional predicates.

3.2. Lewis's Revised Conditional Analysis

Lewis (1997) develops his Revised Conditional Analysis (RCA) in response to Martin's fink cases. While the full articulation of the analysis is quite complex, the basic idea is relatively simple. According to Lewis, to have a disposition is to have a causal basis for that disposition, where a causal basis of a disposition is some property of the object that would be causally responsible for disposition's manifestation if the circumstances of manifestation were to occur. For example, when a fragile glass is struck and consequently breaks, some structural property of the glass is causally responsible for the glass breaking, and this property is the causal basis of the glass's fragility (149). What finks do, then, is destroy the disposition's causal basis, and what reverse finks do is change the object so that it acquires a disposition's causal basis. Lewis's (157) account stipulates that the object has the disposition when and only when it has a causal basis for that disposition. Consequently, RCA is not falsified by finks, the final formulation of which runs as follows: (RCA) Something x is disposed at time t to give response r to stimulus s iff, for some intrinsic property B that x has at t, for some time t′ after t, if x were to undergo stimulus s at time t and retain property B until t′, s and x's having of B would jointly be an x-complete cause of x's giving response r, where an 'x-complete cause of x's giving response r' is a complete cause of x's giving r in-so-far-as x's intrinsic properties are concerned. Since the stimulus s is said to be part of the x-complete cause of r, and the x-complete cause of r only concerns x's intrinsic properties, RCA has the odd consequence that the stimulus for a disposition is counted as an intrinsic property of the disposed object. Perhaps that is a technical problem that can be remedied with careful rewording. But RCA raises other issues that are more central to the prospects of the reductive project.

Note that, according to RCA, the intrinsic property B would be a cause of a certain effect in certain circumstances. Without further analysis of what it means to say that a property would be a cause (see McKitrick 2009 for a characterization of how Lewis could give such an analysis), RCA's *analysans* includes a property which apparently has causal powers. Furthermore, it is also instructive to note what RCA does not do. First, RCA does not say what dispositions *are*, since Lewis (1997, 151) is neutral on the question of whether dispositions are to be identified with their causal basis, or with the second-order property of having some causal basis or other. Second, RCA does not rule out the possibility of fundamental causal powers. Despite the fact that Lewis (148) deems Martin's theory of irreducible dispositionality 'radical', he also writes, 'let us agree to set aside baseless dispositions, if such there be. Our goal, for now, is a reformed conditional analysis of based dispositions' (149). Third, RCA does not reduce ordinary dispositional predicates such as 'fragility' or 'solubility'. Lewis (153) acknowledges 'the first problem we face in analysing any particular dispositional concept, before we can turn to the more general questions...is the problem of specifying the stimulus and the response correctly.' Lewis explains well the challenges that masks and mimics pose for solving this problem. However, he does not take on these challenges, but sets them aside. Focusing on the example of fragility, Lewis writes: 'my purpose in raising this question was not to answer it, but rather to insist that it is merely the question of which response-specification is built into the particular dispositional concept of fragility.... it affords no lesson about dispositionality in general' (153). Conceptual analysis of ordinary dispositional predicates is not the project of Lewis's 'Finkish Dispositions'. Lewis seems more interested in a lesson about 'dispositionality in general', but it is not clear what that means. As noted above, the lesson is not so general as to have implications for baseless dispositions, and it remains neutral on what dispositions are. At best, we learn truth conditions for the proposition 'x is disposed at time t to give response r to stimulus s'. More to the point of this chapter, we are given no reason to think that there are no dispositional properties. In fact, Lewis's two proposals for what dispositions are are both views on which dispositions are *properties* that *exist*.

Because Lewis does not try to answer questions about response specifications for dispositional concepts, RCA leaves the vast majority of disposition vocabulary unanalysed and so does not fulfil the reductionist's hope that true disposition ascriptions can be shown to be semantically equivalent to claims that have no apparent reference to dispositional properties. Granted, Lewis's advocacy of Humean supervenience means that he thinks that dispositional properties supervene on intrinsic qualities of point-sized particulars. But in this respect, dispositional properties are no different than most apparently non-dispositional properties, such as shapes. Lewis's belief that dispositional properties supervene on something merely shows that the concept of a disposition that he analyses is a concept of a non-fundamental property. So, even within the neo-Humean tradition, the suspicion that dispositions are 'occult' and 'ethereal' is diminished to the mere assertion that dispositions are non-fundamental properties. While Humean supervenience implies that there are no necessary connections between distinct existences in the actual world, according to RCA, when something has a disposition, a certain counterfactual is true of it—there is a modal connection between having a disposition and exhibiting a manifestation. It's just that, for Lewis, the source of this modality is not the object itself, but other possible worlds.

3.3. Manley and Wasserman's Proportionality Account

Manley and Wasserman (2007; 2008) add to the pile of problems for previous counterfactual analyses, including SCA and RCA. In addition to posing their Achilles' heel and reverse Achilles' heel counterexamples, Manley and Wasserman point out features of disposition ascriptions that other counterfactual analyses fail to account for. One observation is that some dispositions seem to have no stimulus conditions. For example, some loquacious people are just prone to talk sometimes for no particular reason. Consequently, it is not clear what the content of the antecedent of the counterfactual could be (Manley and Wasserman 2008, 72).[3]

Other problems for previous counterfactual analyses that Manley and Wasserman raise start with a cluster of interrelated observations about disposition ascriptions. First, they are context-sensitive. A wooden beam that is called 'fragile' on a construction site would not be called 'fragile' in an antique shop. Any analysis that specifies one stimulus-manifestation pair to correspond to 'fragile' in one context renders the wrong result in other contexts.

---

[3] Vetter (2014) makes a similar argument, but argues for the stronger conclusion that no dispositions have stimulus conditions.

Second, disposition ascriptions can be comparative. A wine glass is more fragile than a coffee mug. Any analysis that gives the same analysis of the glass's fragility and the mug's fragility cannot do justice to such assertions. Third, the applicability of a dispositional predicate can be a matter of degree, or as Manley and Wasserman put it, many dispositional predicates are 'gradable'. Some things are 'extremely fragile' while other things are 'somewhat fragile'. However, the counterfactuals invoked by conditional analyses do not admit of degree. If a counterfactual analysis is to translate all dispositional vocabulary in terms of pairs of circumstances and manifestations, it would have to find a pair for every degree of fragility, as well as pairs for every degree of every other gradable dispositional predicate.

Remarkably, Manley and Wasserman (76) propose an alternative that promises to avoid all of these problems. They begin by loosening the restriction that, in order to have a certain disposition, one must manifest that disposition whenever the relevant circumstances obtain. A thing can have a disposition even if there are some occasions when it does not manifest that disposition in the circumstances—as long as it does so on enough of those occasions, or 'some suitable proportion'. This is the idea behind their proportionality account, which they state as follows: (PROP) N is disposed to M when C if and only if N would M in some suitable proportion of C–cases. A 'C–case' is a 'stimulus condition case'—a fully specific scenario that settles everything causally relevant to the manifestation of the disposition. Masks, mimics, and finks do not appear to defeat this account since C–cases that include them would presumably be rare enough that the proportionality claim would still be true. Dispositions without stimuli are not a problem, since the C–cases do not need to be restricted to situations in which a certain kind of stimulus occurs. Furthermore, since the suitability of a proportion is a contextually determined matter of degree, the account nicely deals with the fact that dispositional predicates are context-sensitive and gradable.

Before assessing PROP, a few clarifications are in order. First, despite the fact that no explicit counterfactual statement appears in PROP, Manley and Wasserman (2011, 1194 n7) do consider PROP to be a counterfactual analysis. They write: 'PROP... requires, in effect, that [in order to have a disposition] one satisfy a suitably high proportion of a long list of conditionals involving very precise stimulus conditions.' Second, PROP is not an analysis of ordinary dispositional predicates such as 'fragile', 'soluble', or 'flexible'. Manley and Wasserman (2011, 1209) say, 'PROP only concerns explicit dispositional predicates' such as 'disposed to break when dropped'. Third, while Manley and Wasserman claim to be neutral about the reductive project that is the subject of this chapter, they think that their account is capable of providing such a reduction. They write: 'We take PROP to capture a necessary connection between dispositions and conditionals, but we withhold judgment on the priority of either side of this equation (whether metaphysical or conceptual). For that reason, we are agnostic about whether or not PROP counts as a genuine *analysis*. But we are certainly happy to recommend PROP to those who seek such a reduction' (1194).

So, how effective is PROP at semantically reducing dispositional predicates? As noted above, it does not reduce ordinary dispositional predicates. In order to do that, one would have to supplement PROP with translations of ordinary dispositional predicates, such as 'fragile', into explicit dispositional predicates such as 'is disposed to break when dropped'. So, the analysis of ordinary dispositional predicates, such as 'fragile', would proceed by two steps:

1. '*x* is fragile' iff '*x* is disposed to break when dropped'; and
2. '*x* is disposed to break when dropped' iff '*x* would break in some suitable proportion of dropping-cases.'

PROP only concerns the second step. In order to complete the first step, as Lewis notes, one would have to correctly specify the stimulus and response associated with the given dispositional concept. If there are things that we call 'fragile' that don't break when dropped, but instead break in other types of circumstances, such as exposure to heat (Vetter 2015, 41), then (1) is false and, consequently, even if (2) is true, the right side of (2) will not be an analysis of 'fragile'.

Perhaps stimulus and response specifications for every ordinary dispositional predicate can be determined, and the reductionist just has a lot of work to do. But there do seem to be some recalcitrant counterexamples. Consider again the shy, intuitive chameleon who is disposed to look green when someone looks at him but loses that disposition when someone looks at him. Cases where he is looked at and looks green will be extremely rare, and so they will not be a suitable portion of the cases where he is looked at. Since his being shy and intuitive are intrinsic properties of the chameleon, those features that fink his disposition to look green would travel with him, so to speak, across the various C–cases. This is a case of an intrinsic fink (Bird forthcoming), and it is not clear if PROP renders the correct verdict when it comes to intrinsic finks.

The more substantial worry for the reductionist is the nature of the C–cases mentioned in PROP. A C–case is a possible scenario, and for every explicit dispositional predicate there are infinitely many of them; as Manley and

Wasserman state, 'the relevant proportions will have to be achieved by measures on infinite sets' (Manley and Wasserman 2011, 1215 n44). So, PROP makes apparent reference to infinitely many possible scenarios. So much for Goodman's (1979, 35) desire to avoid introducing 'any such troublesome word as "possible"'. PROP relies on the Lewis–Stalnaker semantics for counterfactuals, so, as suggested earlier, the price of reducing disposition ascriptions might be a multiverse of possible worlds. And the account is more complex than it might first appear. Manley and Wasserman (2008, 79) note that they cannot treat all C–cases equally but must weigh some more heavily than others. Furthermore, since there are infinitely many C–cases for any explicit dispositional predicate, the comparative judgements that Manley and Wasserman claim that PROP provides require comparing relative sizes of infinite sets. So, 'the glass is more disposed to break when dropped than the mug is' is true if and only if the glass would break in more dropping-cases than the mug would (77). But since {the dropping-cases in which the glass would break} and {the dropping-cases in which the mug would break} are each infinite in size, it is difficult to see by what standard we should determine which set is larger. Even if we could fill out the complex details of how to appropriately weigh and compare infinitely large sets of C–cases, those complex details are out of reach for those who competently make disposition ascriptions, rendering claims of semantic equivalence problematic (Choi and Fara 2012).

4. Beyond Semantic Reduction

Some reductionists may hold out hope that PROP or some other method of semantic reduction will ultimately work. But after eighty years of trying, hope is wearing thin. One may feel, as Nelson Goodman did in 1953, that we've been beating our heads against a wall and chasing down blind alleys, and it is time to try something different.

4.1. Reductionist Responses

A reductionist can claim that, even though dispositional predicates are not semantically reducible to non-dispositional predicates, dispositions are nevertheless *metaphysically* reducible to non-dispositional entities. There are a couple of ways this suggestion can be understood. One way is that there are dispositional properties which stand in the reduction relation to non-dispositional properties. But notice, if that is the claim, then it is acknowledged that there are dispositional properties. That, by itself, takes them out of the dustbin of history. If dispositions are reducible in this sense, then that just shows that they are not fundamental. There are many non-fundamental things that do not belong in the dustbin of philosophical history—knowledge, belief, and language, to name a few. If dispositions exist, then the debate can turn to whether any of them are metaphysically fundamental.

The other way to interpret the claim that dispositions are metaphysically reducible to non-dispositional entities, attributable to D. M. Armstrong (1996, 18), is that true disposition ascriptions have categorical (aka. non-dispositional) truthmakers. Dispositional properties are not merely reduced but eliminated. Armstrong (17) acknowledges that he 'owes an account of why we are nevertheless entitled to attribute unrealized powers, potentialities, and dispositions to the objects.' To provide such an account, Armstrong appeals to laws of nature: 'The idea is this: given the state of the glass, including its microstructure, plus what is contrary to fact—that the glass is suitably struck—then, *given the laws of nature as they are*, it follows that the glass shatters.... This is what it *is* for the glass to be brittle, and it does not involve anything but categorical properties of the glass.' Such a view needs an account of laws of nature, which Armstrong (1983) provides. On Armstrong's view, particulars instantiate a sparse number of categorical Universals. Some of these Universals are related by a higher-order Universal—the Necessitation relation. When two Universals F and G are so related, it is a natural law: $N (F, G)$.

Since Armstrong's view of dispositions depends on his account of laws, which in turn depends on his account of Universals, his metaphysical reduction of dispositions is only as successful as his larger metaphysical system, the assessment of which would take us too far afield.[4] But note that Armstrong's Universals are immanent, that is, located where they are instantiated. So, consider a particular x that is F, where properties F and G stand in the Necessitation relation. The Necessitation relation is located where x is. Hence, it seems that x is intrinsically necessitated to yield G, and this seems similar to x having a power to produce G. Also, notice that Armstrong's statement of how disposition ascriptions relate to categorical properties is a *counterfactual* statement, so it may not be an alternative to counterfactual analyses after all. In fact, we can interpret the above quote as committing him to 'Armstrong's

---

[4] See Rodriguez-Pereyra (2015) for criticisms of Armstrong's realism and Lewis (1983) and van Fraassen (1987) for criticisms of his account of laws.

Conditional Analysis': (ACA) x is fragile iff [given that x has a certain microstructure, if x were struck, then (given the laws of nature, it follows that x shatters)]. However, we now know that such a biconditional cannot be true and attempts to amend it will take us down the rabbit hole once again.

Armstrong (1996, 40) has another proposal about dispositional predicates which he does not distinguish from the one above. It is that a dispositional predicate is an often-useful way of denoting its causal basis, where this causal basis is a categorical property. This proposal would show that dispositional predicates have semantic values, regardless of the success of any conceptual analysis of their meaning. However, the proposal has certain shortcomings. One is that Armstrong (1989b, 87) believes that the only properties are a sparse number of perfectly natural Universals, so that *most* predicates do not refer to properties. Unless some perfectly natural property is the causal basis of fragility, 'fragile' does not denote a property. If calling something 'fragile' is not a way of talking about some categorical property, and ACA is false, it is not clear what sense has been made of disposition ascriptions. Furthermore, even if some instance of fragility has a genuine property as its causal basis, other instances of fragility have different causal bases. It follows, on this view, that we speak falsely when we say that an eggshell and a wine glass are both 'fragile'. At best, they have different 'fragilities'. While Armstrong says that he owes us an account of justified disposition ascriptions, the ones he provides are problematic.

So, how does one make sense of our ubiquitous use of dispositional predicates if dispositions do not exist? One might insist that true disposition ascriptions are somehow made true by non-dispositional entities even though semantic reduction of dispositional predicates is not possible. But in order for some powerless entity to make some disposition ascription true, it would have to be supplemented with something like natural laws, or some general story about the source of modality. Without some account of these non-dispositional truth-making entities and how they make true disposition ascriptions true, commitment to metaphysical elimination of dispositions looks like mere neo-Humean faith in the idea that powerless entities explain why things seem to have powers.

4.2. Anti-reductionist Responses

The apparent failure of semantic reduction of dispositional predicates motivates other philosophers to reject reduction. George Molnar (2003, 21) calls for 'less conceptual analysis, more metaphysics'. C. B. Martin well articulates this anti-reductionist response when he writes: 'if, as we have *seen*, counterfactuality or strong conditionality cannot explain dispositions, then there is no place to turn but to actual first-order dispositions or powers' (Martin 1994, 7). Stephen Mumford (2009, 170) takes the failure of the Naïve Conditional Analysis, and the need for actual-world truthmakers for stronger conditionals, to be evidence for powers: 'One might then say in response that the conditional would have to be stronger than material. But that looks like a conditional with some modal power and it might then be wondered how that itself can be empirically known. What, in other words, would be an empirically acceptable truthmaker for any such stronger-than-material conditional? The realist about powers has an answer: the power is the truthmaker for any such conditional.' Now that powers are out of the dustbin, they populate ontologies and accomplish many things. Some (Ellis 2001; Bird 2007; Borghini and Williams 2008; Mumford and Anjum 2011; Vetter 2015) claim that dispositions are fundamental, and ground other features of our ontology, including causation, laws of nature, and possibility. Barbara Vetter (2015, 2) asks us to recognize 'potentiality as an *explanans* in the metaphysics of modality, rather than as something in need of explanation and reduction'. Dispositions are thought to be properties with causal 'oomph', likened to Newtonian forces (Schrenk 2009, 164). Dispositional essentialists (Ellis 2001; Bird 2007) claim that *all* natural properties have a dispositional essence. Advocates of the 'dual aspect view' (Martin 1997; Heil 2003) hold that all properties have both dispositional and qualitative aspects. Pandispositionalists or dispositional monists (Shoemaker 1980; Mumford 2004; Bird 2007; Whittle 2008; Borghini and Williams 2008) claim that all properties are dispositions. Causal structuralists (Hawthorne 2001) claim that the identity of every property is determined by its place in a causal structure. Furthermore, in addition to their observable effects, powers are thought to have invisible manifestations that contribute to those effects (Molnar 2003; Cartwright 2009; Corry 2009). Furthermore, powers are thought to be always manifesting, though this is often indiscernible because they counteract each other (Mumford and Anjum 2011; Vetter 2014).

But perhaps there has been an overreaction to the failure of the reductive project. If the semantic irreducibility of dispositional predicates means that those predicates denote powers, then every time an irreducible dispositional predicate is ascribable to something, we should think that thing has a powerful property. So, some book that is readable has, as one of its properties, a property with causal oomph to bring about someone reading it, and if a proposition is debatable, the proposition instantiates a causal power to make it the case that it is the subject of a debate. However, many powers theorists (Hawthorne 2001; Molnar 2003; Bird 2007) want to restrict their claims to natural properties. If the neo-Aristotelian does not want to attribute causal powers to things in virtue of being aptly called 'readable' or 'debatable', and those predicates are no more amenable to analysis than any others, then an inability to reduce a true disposition ascription is not sufficient evidence for the existence of a powerful property.

On my view, which I can only briefly characterize here, there is still an important connection between dispositions and counterfactuals. A thing's disposition is its property of having a certain kind of counterfactual hold of it. However, our language is vague and inexact, and so we do not precisely articulate these counterfactuals with our disposition ascriptions. Rather, to ascribe a dispositional predicate to a thing is to roughly indicate some counterfactual fact that holds of it. These counterfactuals are of a certain type, whereby a certain kind of manifestation would occur if the thing were in certain kinds of circumstances. Such counterfactual claims have various kinds of truthmakers, including non-dispositional properties, dispositional properties, relations, and historical properties. However, it is possible that some counterfactuals hold of a thing because it has a power.

I make a distinction between fundamental dispositions or powers and non-fundamental, derivative dispositions (as does Bird 2007). Like derivative dispositions, powers are such that, when an object has one, a certain kind of counterfactual holds of it. But whereas derivative dispositions are associated with counterfactual claims that can hold for any number of reasons, when it comes to a power, the counterfactual that holds of the object with the power is made true, or is grounded by, the fact that the object has that power. If something has a power, then the correct answer to the question 'in virtue of what does it have this power?' is: 'Not in virtue of anything else. That's just the way that it is.' What I call 'powers' have been called 'pure powers', 'potencies', 'ungrounded dispositions', 'pure dispositions', and 'bare dispositions' (Johnston 1992; McKitrick 2003; Mumford 2006; Bird 2007; Bauer 2012).

So, among the dispositions, which are derivative and which are powers? How do we know when a counterfactual is grounded in the fact that something has power rather than in the fact that it has another derivative disposition or a non-dispositional property? Often, we don't. Knowledge of fundamentality is not easy to come by. But then why think that there are any powers, given that we lack conclusive evidence that any given disposition is a power? Given a true counterfactual of the relevant type, there are four options with respect to its being grounded:

1. It states an ungrounded, fundamental fact.
2. It states a fact that is grounded by other counterfactual facts, which are in turn grounded by other counterfactual facts, ad infinitum.
3. It states a fact that is ultimately and fully grounded in facts about fundamental, non-dispositional entities.
4. It states a fact that is ultimately, and at least partially, grounded in facts about fundamental, dispositional entities. That is to say, it is grounded in something's having a power.

If (4) is a viable option, there is reason to think that some dispositions are powers, even if we are not certain which ones.

References

Armstrong, D. M. 1973. *Belief, Truth and Knowledge*. Cambridge: Cambridge University Press.

Armstrong, D. M. 1983. *What is a Law of Nature?* Cambridge: Cambridge University Press.

Armstrong, D. M. 1989a. *A Combinatorial Theory of Possibility*. Cambridge: Cambridge University Press.

Armstrong, D. M. 1989b. *Universals: An Opinionated Introduction*. Boulder: Westview Press.

Armstrong, D. M., C. B. Martin, and Ullin T Place. 1996. *Dispositions: A Debate*. London: Routledge.

Ayer, A. J. 1936. *Language Truth, and Logic*. London: V. Gollancz.

Bauer, William. 2012. 'Four Theories of Pure Dispositions'. In *Properties, Powers and Structures: Issues in the Metaphysics of Realism*, ed. Alexander Bird, Brian Ellis, and Howard Sankey, xx. New York: Routledge.

Bird, Alexander. 1998. 'Dispositions and Antidotes'. *The Philosophical Quarterly* 48, no. 191 (April): 227–34.

Bird, Alexander. 2007. *Nature's Metaphysics: Laws and Properties*. Oxford: Oxford University Press.

Bird, Alexander. Forthcoming. 'Can Dispositions have Intrinsic Finks and Antidotes?' (manuscript).

Block, Ned. 1990. 'Can the Mind Change the World?' In *Meaning and Method: Essays in Honor of Hilary Putnam*, ed. George Boolos, 137–70. Cambridge: Cambridge University Press.

Borghini, Andrea, and Neil E. Williams. 2008. 'A Dispositional Theory of Possibility'. *Dialectica* 62, no. 1 (January): 21–41.

Carnap, Rudolf. 1936. 'Testability and Meaning'. *Philosophy of Science* 3, no. 4 (October): 419–71.

Cartwright, Nancy. 2009. 'Causal Laws, Policy Predictions, and the Need for Genuine Powers'. In Handfield 2009, 127–57.

Choi, Sungho. 2006. 'The Simple vs. Reformed Conditional Analysis of Dispositions'. *Synthese* 148, no. 1 (January): 369–79.

Choi, Sungho. 2008. 'Dispositional Properties and Counterfactual Conditionals'. *Mind* 117, no. 468 (October): 795–841.

Choi, Sungho, and Michael Fara. 2012. 'Dispositions'. In *Stanford Encyclopedia of Philosophy*, rev. 5 January 2012. https://plato.stanford.edu/archives/spr2016/entries/dispositions/.

Corry, Richard. 2009. 'How is Scientific Analysis Possible?' In Handfield 2009, 158–88.

Cross, Troy. 2005. 'What is a Disposition?' *Synthese* 144, no. 3 (April): 321–41.

Damschen, Gregor, Robert Schnepf, and Karsten R. Stüber, eds. 2009. *Debating Dispositions: Issues in Metaphysics, Epistemology and Philosophy of Mind*. Berlin: Walter de Gruyter.

Ellis, Brian. 2001. *Scientific Essentialism*. Cambridge: Cambridge University Press.

Goodman, Nelson. 1979. *Fact, Fiction, and Forecast*. 3rd ed. Cambridge: Harvard University Press.

Gundersen, Lars. 2002. 'In Defense of the Conditional Account of Dispositions'. *Synthese* 130, no. 3 (March): 389–411.

Handfield, Toby, ed. 2009. *Dispositions and Causes*. Oxford: Clarendon Press.

Hauska, Jan. 2008. 'Dispositions and Normal Conditions'. *Philosophical Studies* 139, no. 2 (May): 219–32.

Hawthorne, John. 2001. 'Causal Structuralism'. *Philosophical Perspectives* 15: 361–78.

Heil, John. 2003. *From an Ontological Point of View*, Oxford: Clarendon Press.

Jackson, Frank. 1995. 'Mental Properties, Essentialism and Causation'. *Proceedings of the Aristotelian Society*, n.s., 95: 253–68.

Johnston, Mark. 1992. 'How to Speak of the Colors'. *Philosophical Studies* 68, no. 3 (December): 221–63.

Kaila, Eino. 1939. *Inhimillinen tieto, mita" se on ja mita" se ei ole* ('Human Knowledge. What it is and what it is not'). Helsingfors: Otava.

Kaila, Eino. 1941. 'Über den physikalischen Realitätsbegriff: Zweiter Beitrag zum logischen Empirismus'. *Acta philosophica fennica* 4: 33–34.

Lewis, David. 1973. *Counterfactuals*. Cambridge: Harvard University Press.

Lewis, David. 1979. 'Counterfactual Dependence and Time's Arrow'. *Noûs* 13: 455–76.

Lewis, David. 1983. 'New Work for a Theory of Universals'. *Australasian Journal of Philosophy* 61 (4): 343–77.

Lewis, David. 1983–6. *Philosophical Papers*. 2 vols. Oxford: Oxford University Press.

Lewis, David. 1986. *On the Plurality of Worlds*. Oxford: Blackwell.

Lewis, David. 1997. 'Finkish Dispositions'. *The Philosophical Quarterly* 47, no. 187 (April): 143–58.

McGlynn, Aidan. 2012. 'The Problem of True-True Counterfactuals'. *Analysis* 72, no. 2 (April): 276–85.

McKitrick, Jennifer. 2003. 'The Bare Metaphysical Possibility of Bare Dispositions'. *Philosophy and Phenomenological Research* 66, no. 2 (March): 349–69.

McKitrick, Jennifer. 2009. 'Dispositions, Causes, and Reduction'. In Handfield 2009, 31–64.

Malzkorn, Wolfgang. 2000. 'Realism, Functionalism and the Conditional Analysis of Dispositions'. *The Philosophical Quarterly* 50, no. 201 (October): 452–69.

Malzkorn, Wolfgang. 2001. 'Defining Disposition Concepts: A Brief History of the Problem'. *Studies in History and Philosophy of Science Part A* 32, no. 2 (June): 335–53.

Manley, David, and Ryan Wasserman. 2007. 'A Gradable Approach to Dispositions'. *The Philosophical Quarterly* 57, no. 226 (January): 68–75.

Manley, David, and Ryan Wasserman. 2008 'On Linking Dispositions and Conditionals'. *Mind* 117, no. 465 (January): 59–84.

Manley, David, and Ryan Wasserman. 2011. 'Dispositions, Conditionals, and Counterexamples'. *Mind* 120, no. 480 (October): 1191–227.

Martin, C. B. 1994. 'Dispositions and Conditionals' *The Philosophical Quarterly*, 44, no. 174 (January): 1–8.

Martin, C. B. 1997. 'On the Need for Properties: The Road to Pythagoreanism and Back'. *Synthese* 112, no. 2 (August): 193–231.

Mellor, Hugh. 2000. 'The Semantics and Ontology of Dispositions'. *Mind* 109, no. 436 (October): 757–80.

Molnar, George. 2003. *Powers: A Study in Metaphysics*, Oxford: Oxford University Press.

Mumford, Stephen. 1998. *Dispositions*. Oxford: Clarendon Press.

Mumford, Stephen. 2004. *Laws in Nature*. New York: Routledge.

Mumford, Stephen. 2006. 'The Ungrounded Argument'. *Synthese* 149, no. 3 (April): 471–89.

Mumford, Stephen. 2009. 'Ascribing Dispositions'. In Damschen et al. 2009, 168–85.

Mumford, Stephen, and Rani Lill Anjum. 2011. *Getting Causes from Powers*. Oxford: Oxford University Press.

Pap, Arthur. 1958. 'Disposition Concepts and Extensional Logic'. In *Concepts, Theories, and the Mind-Body Problem*, Minnesota Studies in the Philosophy of Science 2, 196–224. Minneapolis: University of Minnesota Press.

Prior, Elizabeth. 1985. *Dispositions*. Aberdeen: Aberdeen University Press.

Rodriguez-Pereyra, Gonzalo. 2015. 'Nominalism in Metaphysics'. In *Stanford Encyclopedia of Philosophy*, rev. 1 April 2015. https://plato.stanford.edu/entries/nominalism-metaphysics/.

Rosen, Gideon. 1990. 'Modal Fictionalism'. *Mind* 99, no 395 (July): 327–54.

Ryle, Gilbert. 1949. *The Concept of Mind*. London: Penguin.

Scholz, Oliver R. 2009. 'From Ordinary Language to the Metaphysics of Dispositions: Gilbert Ryle on Dispositions Talk and Dispositions'. In Damschen et al. 2009, 127–44.

Schrenk, Markus. 2009. 'Hic Rhodos, Hic Salta: From Reductionist Semantics to a Realist Ontology of Forceful Dispositions'. In Damschen et al. 2009, 145–67.

Shoemaker, Sydney. 1980. 'Causality and Properties'. In *Time and Cause: Essays Presented to Richard Taylor*, ed. Peter van Inwagen, 109–35. Dordrecht: Reidel.

Shoemaker, Sydney. 2001. 'Realization and Mental Causation'. In *The Proceedings of the Twentieth World Congress of Philosophy*, ed. Carl Gillett and Barry M. Loewer, 23–33. Cambridge: Cambridge University Press.

Smith, A. D. 1977. 'Dispositional Properties'. *Mind* 86, no. 343 (July): 439–45.

Stalnaker, Robert. 1968. 'A Theory of Conditionals'. In *Studies in Logical Theory*, ed. Nicholas Rescher, 98–112. Oxford: Blackwell.

Stalnaker, Robert. 1976. 'Possible Worlds'. *Noûs* 10:65–75.

Steinberg, Jesse. 2010. 'Dispositions and Subjunctives'. *Philosophical Studies* 148 no. 3 (April): 323–41.

Storer, Thomas. 1951. 'On Defining "Soluble"'. *Analysis* 11, no. 6 (June), 134–7.

Tornaletti, Sylvia, and Gerd P. Pfeifer. 1996. 'UV Damage and Repair Mechanisms in Mammalian Cells'. *BioEssays* 18, no. 3 (March): 221–28.

Van Fraassen, Bas. 1987. 'Armstrong on Laws and Probabilities'. *Australasian Journal of Philosophy*, 65 (3): 243–59.

Vetter, Barbara. 2014. 'Dispositions without Conditionals'. *Mind* 123, no. 489 (January): 129–56.

Vetter, Barbara. 2015. *Potentiality: From Dispositions to Modality*. Oxford: Oxford University Press.

Whittle, Ann. 2008. 'A Functionalist Theory of Properties'. *Philosophy and Phenomenological Research* 77, no. 1 (July): 59–82.

11

Causal Powers and Structures

Brian Ellis

In this chapter it will be argued that causal powers and structures are two essentially different kinds of properties. The first is dispositional, and the second categorical. It will also be argued that both kinds of properties are required for an adequate scientifically realistic account of the nature of reality.

1. Metaphysical Background

Recent developments in metaphysics have seen a strong revival of the idea that things of different natural kinds are distinguished essentially by their powers and structures. In *Scientific Essentialism* (Ellis 2001), I argued that the world must be thought to be made up of hierarchies of natural kinds of objects whose behaviours are linked by natural kinds of processes. I was inspired initially by the structure of chemistry. To make good sense of all that had been discovered in this field, I argued that comprehensive hierarchies of natural kinds of substances and processes of chemical combination must exist.

Firstly, we must suppose that there are more than ninety mutually distinct chemical elements occurring naturally in the world, and literally hundreds of thousands of chemical compounds of these elements, all of which are categorically distinct from one another. These compounds are not mixtures, which could be in any proportions; they are discrete and in definite proportions. The question then becomes how to explain all this variety of discrete kinds of substances. Each substance, I argued, must be supposed to be essentially different from every other, because it is manifestly not true that there is a continuum of chemical variety. Moreover, the chemical elements and compounds must not only be categorically distinct from one another but also fixed in their natures. That is, they must all be members of distinct natural kinds. For obvious reasons, I called these substantive kinds, and argued that substantive kinds had all of the characteristics of Aristotelian universals.

Brian Ellis, *Causal Powers and Structures* In: *Reconsidering Causal Powers: Historical and Conceptual Perspectives*. Edited by: Benjamin Hill, Henrik Lagerlund, and Stathis Psillos, Oxford University Press (2021). © Brian Ellis.
DOI: 10.1093/oso/9780198869528.003.0012

Secondly, we must suppose that the chemical processes described in detail in chemistry textbooks are all essentially different from one another, since each is represented by a different chemical equation. Moreover, the processes of chemical combination and de-combination are all causal processes, and in many cases they can be driven backwards or forwards depending upon the temperature or other conditions in which the relevant ingredients exist. Therefore, to explain all of this variety and discreteness in the chemosphere we must suppose that there are literally hundreds of thousands of natural kinds of chemical processes, each distinct from every other in its nature. I called the natural kinds of processes 'dynamic kinds', and argued that these too had all of the characteristics of classical universals.

The variety of chemical compounds can plausibly be explained simply as just some of the different possible discrete structures of the elements of which they are composed. But as we descend the scale of molecular complexity down to the chemical elements, we are forced to look for the sources of their differences at a deeper level, the subatomic one. The ninety-odd mutually distinct, naturally occurring chemical elements may perhaps be distinguished by a corresponding number of different kinds of atoms, with differently structured nuclei and different electron structures surrounding them.

We know that there are structures, because different structures are needed to explain the existence of isomers, that is molecules of chemically distinct substances having physically distinct X-ray profiles but the same molecular formulae. Moreover, the postulated structures not only explain precisely the different X-ray diffraction patterns that various isomers have; they also enable chemists to build new molecular structures out of the known ones and predict what their causal powers are likely to be. Indeed, if we did not have the kind of knowledge that X-ray diffraction technology has yielded, modern genetics and biochemistry would both be impossible. If you believe in the double helix structure of all animal genomes, then you must believe in the fundamental existence of physical and chemical structures in the natural world.

However, this process of explaining the distinctions between the kinds by distinguishing their structures has its limitations. For, sooner or later, we must run out of structure. Therefore, we must eventually come down to things that cannot be distinguished by their structures. Let us just call these things the 'fundamental particles', and leave it an open question what these fundamental particles are. The question then arises: how are the fundamental particles to be distinguished from one another? The only plausible answer is that they must be distinguished by their properties. But their properties, whatever they might be, cannot be structures. For, if the fundamental particles were structured, they would not be fundamental. Therefore, the properties of the fundamental particles must be simple, that is ones that are distinguished from one another, not by what they are, as structures are distinguishable, but by what they dispose their bearers to do. That is, they must be dispositional properties, just as causal powers are supposed to be. Therefore, to make sense of all this evident structure in the world, we must suppose that, at this basic level, things are distinguished, not by what they are, but by how they are disposed to behave.

The dispositional properties of things can be either causal powers or propensities. Their causal powers would ground the ways they are disposed to act on, or react to, other things. And their propensities would ground the ways, and the probabilities with which, they are intrinsically disposed to change themselves spontaneously, independently of other things. Therefore, at the level of subatomic particles, we must suppose that things are distinguished from one another by these two sorts of properties, causal powers and propensities. Scientific realism thus implies a sort of natural-kind realism, which in turn implies realism about causal powers and propensities all the way down to the fundamental particles.

The causal powers and propensities of things are postulated to be a species of classical universal, the displays of which are, respectively, natural kinds of causal processes, similar to those that we encounter in chemistry laboratories, and natural kinds of decay processes such as those we observe in Wilson cloud chambers. Therefore, a causal power must be understood to be a primitive relationship of possession between an object (that is a substance) and a specific kind of universal that disposes its bearer to participate in a display of that causal power (in appropriate circumstances). Formally, this explanation is no different from Aristotle's. An object is blue, according to Aristotle, if and only if it stands in the relationship of possessing the disposition to look blue in certain standard kinds of circumstances. Aristotle's universals are thus tropic universals, the instances of which are the tropes of the properties they seek to explain. The same is true of causal powers and propensities; their instances are their dispositions to participate in these natural kinds of processes in appropriate circumstances.

The causal powers and propensities of the fundamental particles in nature cannot be functions of their relations to other things. For, if they were, the identities of the particles would be different in different locations. But they are not. Therefore, the dispositional properties that determine the identities of things all the way down to the level of the fundamental particles must be intrinsic to them. These identity-determining intrinsic properties are, in

Locke's sense, the real essences of the things that bear them. They are the properties in virtue of which the things in question are things of the kinds they are. The real essence of an electron, for example, must include its mass m, its charge e, and its spin ½. For nothing could be an electron if it lacked any of these properties, and anything that has all of these properties (and whatever other defining properties electrons may have essentially) must be an electron. The real essences of the other fundamental particles of nature, such as protons and neutrons, can likewise be given. In every case, the real essences of the most fundamental particles in nature are sets of dispositional properties; that is, their real essences are their causal powers and propensities (including their half-lives).

This is the theory of the various kinds of properties that I defended in Ellis 2001. It was my answer to Roy Bhaskar's (1975) question: what must the world be like that science as we know it should be possible? I would now argue that, if it is adequate for this much of our scientific understanding of what exists in nature, as I believe it is, then it must have something going for it. It may not be what is ultimately basic in reality. But I do not think that this should matter very much. An ontology that dealt adequately with quantum phenomena, and laid claim to being an account of what exists most fundamentally in nature, would at least have to explain why my ontology of objects with dispositional properties and structures of various natural kinds, and ultimately my account of laws of nature, works as well as it does.

Nevertheless, the theory has been challenged by some philosophers, and I should like to take this occasion to answer a couple of these criticisms that have not, apparently, been answered to everyone's satisfaction.

2. Objections to this Ontology

Firstly, Armstrong (1999) has argued that my causal powers are like human intentions in that they 'point to' their manifestations, even if they are never manifested. I agree that there is a connection between intending and being disposed to act in certain kinds of ways. In fact, being disposed *by choice* to act in a certain kind of way is exactly what it is to intend to do it. But Armstrong's attempt to anthropomorphize my theory is absurd. For it is not true that any object that is disposed to act in a certain kind of way is so by choice. In my recent book *Social Humanism* (Ellis 2012), I argued that every physical object has a causal power profile. But only a very few such objects, viz. some of the higher mammals, have the power to change their causal power profiles deliberately. Their causal power profiles are their dispositions to behave in various ways. In human beings, these dispositions include some dispositions to change (or not to change) the ways in which they are disposed to behave. And this is what it is to be disposed by choice to behave in the way one is. Thus, the exercise of our freedom to choose is the exercise of a meta-causal power.

Meta-causal powers are not all that unusual in nature. In fact, every intelligent animal has the power to learn by experience, which is meta-causal. As I explained (Ellis 2012), every intelligent animal has what I call 'an action profile'. That is, it has a certain set of objectives that has been refined or modified in the light of experience, and a certain range of dispositions, which have been adapted to realizing these ends, and which are to be triggered in circumstances that are appropriate for doing so. Here is what I said:

> An animal's action profile is subtly different from its causal power profile. Its causal power profile is focussed on its dispositions to act or react in specific circumstances. Its action profile, in contrast, is concerned with its intentions, and the courses of action it is disposed to take to realise these intentions. But the action profiles of normal adult humans are superficially very different from those of animals other kinds. The differences are not so great that human beings could not plausibly have developed the capacity to acquire them by natural selection. But many would doubt whether other animals have the capacity to act intentionally. For, intentional actions require having beliefs about what their intended consequences are likely to be, and hence beliefs about probable causation. But surely, it will be said, this is too sophisticated for any non-human creature to grasp. To be sure, human action profiles are extraordinary. The second order causal powers that are required to manipulate our action profiles are not well developed in other animals. There are, no doubt, some such powers displayed by some of the more intelligent animals that we know about, e.g. the anthropoid apes. For, these animals not only have the ability to learn from experience, but also to make some quite sophisticated decisions about how to achieve the things they want, e.g. by the use of primitive tools, such as sticks or stones. Nevertheless, human deliberative powers are in a class of their own, informed as they are by the powerful human capacities of reason, imagination, memory and empathy. We even have a capacity (synderesis) to stand back imaginatively from a particular situation, and view it dispassionately as an individual of no specific identity, and thus come to think about it from an abstract moral point of view.   (76)

In this passage, I have outlined a theory of intentionality based on conceptions of causal power and action profiles. Action profiles, I argued, are meta-causal, and probably exist only in the higher mammals. Consequently, it is absurd to suppose that fundamental particles can have action profiles or engage in intentional activity. The boot is on the other foot. I am able to give a meta-physically sound theory of intentional action, given the existence of causal power and action profiles in human beings. The notion that I am somehow trying to reduce dispositional behaviour to human intentional behaviour is juvenile.

Secondly, the idea that causal powers are primitive relationships of 'having' or 'possessing' certain universals, the instances of which are dispositions to display certain natural kinds of physical processes, seems not to have been properly understood or appreciated. It is not appreciated, I surmise, because universals are widely considered to be just natural kinds of sensory qualities (such as blueness, or sweetness). But I can find no justification in Aristotle's own writings for restricting his theory of universals to just this particular spe-cies of natural kinds. On the contrary, his paradigms of universals include the natural kinds of geometrical shapes. For me, natural kinds of dispositional properties, which are distinguished by the natural kinds of processes they dis-posed to participate in, are universals too.

Thirdly, my distinction between metaphysical necessity and logical neces-sity seems to have fallen on hard ground. Metaphysical necessities are a poste-riori, and so cannot be discovered in the same way as the truths of logic. They are what are normally called 'natural necessities'. Natural necessities, I main-tain, are normally discovered empirically, and have to do with what is physic-ally possible or impossible. Logical necessities are formal necessities and can be discovered by reflecting upon the quantifiers and connectives of (now for-malized) languages, or constructing mathematical arguments based upon such reflections. They are like the truths of mathematics, and independent of experience in the same sort of way. But any language that is adequate for sci-ence must be a modal language, and every such language requires a concep-tion of a possible world, or, alternatively, a rational belief system. So, the question arises: what is the appropriate conception of a possible world for a modal language intended for use in science? In my view, the conception of possibility required for a scientific modal language is just that of metaphysical necessity, and possible worlds must be understood as metaphysically possible worlds. Otherwise, you will simply get the wrong answer when you try to rea-son about what is possible in chemistry. If you restrict your understanding of what is possible to what is physically possible, you will get answers that are consistent with the theories you have about the nature of reality. The modal logics required for science are formally the same as the traditional ones. But the set of metaphysically possible worlds is more restricted than the set of all logically possible worlds.

3. What are Physical Causal Processes?

What we call a 'causal relation' is, typically, a relation between two open physical systems, not a relation between events. One system impacts upon another to cause it to act in some way other than it would, *ceteris paribus*, have acted, if the cause had not been operating. But for one system to impact upon another there must be some physical causal process involved, and typically the impact will involve a transfer of energy of some kind from the impacting body to the body impacted upon. In the case of Hume's billiard ball, for example, some or all of the kinetic energy of the impacting ball is transferred to that of the ball impacted upon. In the act of warming oneself in front of the fire, the chemical energy stored in the wood is released as heat in the process of burning, and the heat energy is transferred radiantly to the person who is warmed. If a tree is blown over in a storm, the force of the wind (that is, its kinetic energy) is partly deflected by the tree on the side exposed to it, and the tree is pushed over by the mechanical force that this deflection generates. Such examples are obviously physical causal processes, and the mechanism of causation is no less obviously one of energy transfer. Therefore, it is initially plausible to suppose that this is really the physical mechanism we are seeking.

It is true that not all physical causal connections seem to be quite like this. In many cases, for example, the causal process is two-way. Thus, electrons repel each other, and gravitational masses attract each other. But it is now widely believed that such causal interactions necessarily involve exchanges of virtual particles—photons in the first of these two cases, and gravitons in the second. So, these interactive cases do not seriously undermine the initially plausible suggestion that physical causation is essentially an energy transfer process. A more serious threat to the energy transfer theory derives from interventions. For we all want to say such things as the eclipse of the sun caused the light to fade, my dark glasses filtered out the ultraviolet light, or the ice cube cooled my drink. And these would all appear to be physical causes and effects. But dimming, shading, filtering out, deflecting, cooling, and so on are all negative causal influences. They affect the system that is influenced by *preventing* an energy transfer process from occurring or having its full effect.

Let us, therefore, distinguish between positively and negatively acting causes, and recognize that most of the processes we call 'causal' involve elements of both. But let us not be too distracted by this. Our natural tendency is to focus on the actions that we may take (for example, shading our eyes), and the consequences that these actions may have for us (for example, reducing the glare). For these are what interest us. But the underlying physical causal process remains the same. It begins with radiation from a light source, and it ends on whatever absorbs or reflects this light. And this remains the case, whatever I may do to shade my eyes. Therefore, the direction of energy transfer remains the same, even though the amount, or nature, of the energy transferred to my eyes is reduced or changed. So, I bite the bullet on this, and say that shading X is not the direct cause of anything that happens on X. It may be preventing something from happening on X. But preventing something from happening is not a way of causing its non-occurrence to happen. Nor is deflecting, filtering out, or any of the other ways of interfering with physical causal processes. The so-called 'negatively acting causes' achieve their effects in very different ways from the positively acting ones. If I shade my eyes, I prevent some or all of the light from reaching them. But this result is brought about by doing something positive, viz. absorbing or reflecting some or all of the light coming towards my eyes. It is not achieved by doing anything to my eyes.

Accordingly, we may define a physical causal process $A \Rightarrow B$ as an energy transference from one physical system S_1 in a state A to another physical system S_2 to effect a physical change B in that system (relative to what the state of S_2 would have been in the absence of this influence). Note that, given this definition, the direction of causation is necessarily that of the energy transfer. Negative causal influences notwithstanding, if there is no energy transfer from A to B, then there is no physical causation in that direction. I have elsewhere (Ellis 2009, 73–92) developed the thesis that all causation consists ultimately of myriads of elementary causal processes in which energy is transmitted by De Broglie, or Schrödinger, waves. But the deep level of ontological analysis required for quantum theory is not an appropriate one for describing the physical causal processes that occur at the object level. At the quantum level, even the physical objects that are the bearers of causal powers have a doubtful ontological status. However, physical objects capable of having causal powers manifestly do exist at what I call 'the object level' of inquiry, that is, the level of ontological reduction at which we may still speak freely of physical objects and their properties.

The physical causal processes that are the displays of causal powers are presumably ones that belong to natural kinds. For, if they were not members of natural kinds, the causal powers would not be natural properties. Nevertheless, there are clearly a great many natural kinds of causal processes in nature, and so, presumably, a great many natural causal powers. All of the chemical reactions described in chemistry books, for example, are processes that belong to natural kinds; and the equations that describe these reactions are presumably descriptive of their essential natures. There are also a great many natural kinds of physical causal processes occurring in other areas of science. So there is no shortage of causal powers.

4. The Laws of Action of the Causal Powers

Every display of a physical causal power requires a bearer from which energy is lost and a receiver to which it is transmitted. Every display of an interactive power requires two sources, one operating in each of the interacting objects, and one receiving in each of these objects. The effectiveness of a power always depends on its magnitude, and usually on the distance over which the energy is transmitted. The displays of elementary causal powers are all natural kinds of processes, and are always law-governed. They accord with the laws I call the 'the laws of action of the causal powers'. For light, the relevant laws are those of electromagnetic radiation. For particle emissions, they are Schrödinger's wave equations. For missiles launched at modest speeds, the transmission laws are the standard Galilean laws of projectile motion. For missiles launched at speeds near to the velocity of light, they are the laws of special relativity. For chemical combinations, or other chemical reactions, the essences of the processes are described by the relevant chemical equations.

In each of these kinds of laws, time and distance are involved essentially, and these variables cannot be eliminated from the laws of transmission of the causal powers. At least, it has not been done by any of the defenders of the view that all properties are essentially dispositional. And, as we shall see, there are very good reasons to think that these paradigmatically categorical properties are involved essentially in the laws of action and transmission of all of the causal powers. Of course, it could be argued that, despite appearances, spatial and temporal relations are essentially dispositional properties. But if this is so, they are neither causal powers nor propensities but properties that are quite unlike any other dispositional properties.

Every instance of a causal power must have some specific location. It must be a causal power of one thing rather than another, or be located in one place rather than another, or be distributed in this region rather than that one. But no instance of location has any specific location contingently. It is just where it is—and necessarily so. A specific instance of any genuine causal power, such as gravitational mass, could be anywhere. Its whereabouts is contingent, and it might have any of infinitely many possible locations. But no specific instance of location has its location contingently. Necessarily, it is where it is. Therefore, location cannot possibly be a normal causal power. Nor is a location a propensity. For there is nothing that a location is intrinsically disposed to do. It might be occupied or vacant. But it is not intrinsically disposed to become one or the other or do anything else that might be said to happen at some point. Moreover, if you remove all of the causal powers from any given location (and this would always seem to be a possibility), the location remains, but the causal powers do not. Therefore, it cannot be said truly that the locations of things have any causal powers essentially.

5. Categorical Realism

If a location is neither a causal power nor a propensity, and locations do not have any causal powers essentially, then what are they? In my view, locations are essentially categorical. According to Heil (2005) and Bird (2005), they must be quiddities, and therefore unreal. And, presumably, any properties or structures that are definable only in the sorts of ways that locations are definable must also be quiddities. But I don't think that locations, or physical or chemical structures, are unreal.

In an earlier work (Ellis 2009, 109), I agreed, somewhat with tongue in cheek, that locations are quiddities, because (a) I did not really mind much what they were called, and (b) I wished to stress that whatever they might be, they were certainly not causal powers. But I did not claim, nor did I think for a moment, that locations have no essential roles in physical causal explanations. On the contrary, I went on to argue that locations, and other categorical properties, are indispensable to explanations invoking causal powers. To cite a causal power by way of explanation of an event is not to explain *that event*, but just to say what kind of explanation it has. To explain a token event, you have to locate both it, and the causal power that allegedly caused it, and show that the event could have been caused by an energy transmission process of the kind that the supposed cause of the action could have provided. My claim was just that locations are not causal powers, and that their roles in causal explanations were nothing like those of causal powers. In fact, I argued specifically that locations are involved essentially in descriptions of *the circumstances* that occasion the displays of causal powers.

Having made this remark, I went on to argue that categorical properties and structures are much more like locations than causal powers, because none of the properties or structures that are definable in terms of spatio-temporal relations would appear to be causal powers. On the contrary, I said, the shapes, sizes, orientations, motions, and so on that are fully definable spatio-temporally all seem to lack the essential properties of powers. None of them has a plausible law of action or produces any effect that might be due to its action. Nor are the categorical properties of things powers to resist, deflect, or otherwise interfere with the actions of any known causal powers, as, for example, elasticity, inertial mass, and electrical resistance clearly are. They would all appear to be just categorical properties, that is properties whose identities depend not on what they do but on what they are. But this is not to say that categorical properties have no roles in the explanation of particular events. Nor is it to say that their roles are restricted to the particular. On the contrary, I think that to explain structures, you need structures. To explain the pattern on a photographic plate, you need a structure of radiant electro-magnetic energy that is adequate to produce that structure. To explain our capacity to see shapes and sizes, things in 3D, and things in perspective, you need structures of things in the visible world adequate to generate these images and provide us with this information. That is, you need to be realists about categorical properties. And, this is what I am—a categorical realist. But I am not exclusively a categorical realist. For I also believe in causal powers and propensities. Therefore, I am also a dispositional realist.

6. Knowledge of Categorical Properties

The fundamentally Lockean position outlined here has been attacked on two fronts. There are strong categoricalists who deny that there are any genuine causal powers, and strong dispositionalists who deny that there are any real categorical properties. I have discussed the principal arguments for and against strong categoricalism elsewhere, and I shall not repeat this discussion here. The main objection to my bipolar position now appears to be coming from strong dispositionalists, who argue that (a) if the categorical properties of things have no causal powers, then we could not know anything about them, and (b) if the causal powers of the different categorical properties were not distinctive of them, then we should have no way of distinguishing between them. Therefore, they say, the categorical properties of things must all have or be distinctive causal powers, that is powers that would distinguish them essentially from one another. Otherwise, it is said, they would all be just quiddities.

This argument from quidditism against the possibility that we could have any knowledge of categorical properties may seem convincing. The argument is prima facie plausible, I suppose, because most philosophers are conditioned to think of causation as a relationship between kinds of events. Hence, if a causal power is to be identified by what it does, the essence of a causal power could only be the kind of event it causes, or at least would cause if it were to be activated. But, for any believer in causal powers, there is a distinction between kinds and tokens of causal powers. Tokens of causal powers are located in the objects involved in token causal processes. But the positions of these objects are not located in these objects. Therefore, the distance between these objects is not a causal power involved in this process. It is, of course, relevant to how long it takes for the energy transmission process between the objects to be completed, and therefore when the effect occurs. And, if the transmission process is radiant, as it is for light or heat, then the distance apart of the objects involved affects the magnitude of the effect it has. Thus, if the transmitter of the causal influence is radiant heat, then the distance apart affects the amount of heat received by the recipient object. The amount of heat received is also a function of the size of the recipient object from the perspective of the heat source. Therefore, apparent cross-sectional area, which is paradigmatically a categorical property, is also an important factor in determining the outcome of the particular causal process.

7. Positive Account of Powers and Structures

Here is my positive account: the dispositional properties of objects are distinguished from one another by their distinctive laws of action. The laws of action for specific properties describe the kinds of effects they have in the kinds of circumstances in which they are placed. The structures of objects are known by the patterns of effects produced by their causal powers. These structures may sometimes be observed directly as our brains process the patterns of our sensory experience. But, in other cases, they must be inferred from the observable patterns detected by our instruments. In these cases, the structure of the objects under investigation can only be deduced, if we know (a) the laws of action of the causal powers involved in their production, and (b) the relevant locations, orientations, and so on, of the objects under investigation, and the powers and circumstances of the instruments employed. For example, to derive the chemical structure of a molecule from an X-ray diffraction pattern we need to know (a) the laws of X-ray diffraction, and (b) the categorical structure of the experimental set-up. Thus, the categorical properties of things, and the categorical circumstances in which they are placed, can never be effective on their own in determining outcomes.

Categorical properties need things with causal powers to be effective. But effective they are. The magnification achieved by a lens, for example, is a function not only of the focal length of the lens, and hence of its magnifying power, but also of its distance from the object that is being magnified. So, distances can be factors in determining outcomes, even though the distances in question are not causal powers.

But have I not, in stating my position, already conceded its falsity? No. In allowing that distances can be factors in determining outcomes in particular cases, have I not conceded that distances can have causal powers? The distances between causally interacting bodies can affect how strongly they interact. But being a factor in determining an outcome is not the same as being a cause or having the causal power to achieve this outcome—at least, not as I have defined physical causation. Living a long way from Sydney is a factor in determining whether or not I can walk there. But it is not a physical causal power, or anything that has such a power. It is also a factor in determining whether or not I can get there by road in less than ten minutes. But it is not a cause of my not being able to get there by road in less than ten minutes. Lengths, orientations, distances, times, shapes, and sizes may all be factors in determining what outcomes are certain, possible, or impossible in particular cases. But this does not imply that they are causal powers, or even that they have any causal powers.

References

Armstrong, D. M. 1999. 'Comment on Ellis'. In *Causation and Laws of Nature*, ed. Howard Sankey, Studies in History and Philosophy of Science 14, 35–38. Dordrecht: Springer.

Bhaskar, Roy. 1975. *A Realist Theory of Science*. Sussex: Harvester.

Bird, Alexander. 2005. 'Laws and Essences'. *Ratio* 18, no. 4 (December): 437–61.

Ellis, Brian. 2001. *Scientific Essentialism*. Cambridge: Cambridge University Press.

Ellis, Brian. 2009. *The Metaphysics of Scientific Realism*. Durham: Acumen.

Ellis, Brian. 2012. *Social Humanism: A New Metaphysics*. New York: Routledge.

Heil, John. 2005. 'Kinds and Essences'. *Ratio* 18, no. 4 (December): 405–19.

12

Induction and Natural Kinds Revisited

Howard Sankey

In this chapter, I revisit a topic about which I wrote more than twenty years ago. In 'Induction and Natural Kinds' (Sankey 1997) I offered a metaphysical solution to the epistemological problem of induction. The solution drew on the insights of Hilary Kornblith (1993) about the relationship between induction and natural kinds. In a departure from Kornblith, I employed Brian Ellis's (2001) account of natural kinds, which I found more plausible than the account of kinds to which Kornblith himself subscribed.

The problem of induction is a problem of circularity. Our spontaneous reaction to the question of how to justify induction is that it has been reliable in the past so will continue to be so in future. But to reason in this way is to employ induction to justify itself, which is to argue in a circle. In my own attempt to justify induction, I employ an inference to the best explanation of the reliability of induction. But inference to best explanation and induction are both forms of ampliative inference. I therefore employ ampliative inference to support ampliative inference. Hence my approach also falls foul of the problem of circularity.

In this chapter, I return to the topic. I take the threat of circularity to be genuine, though perhaps it may be mitigated. I now wish to suggest that there is a way to avoid the circularity. The way to avoid the circularity is to recognize that it is the way the world is that grounds the reliability of induction, not the inference which we employ to establish the grounds of such reliability. My aim in this chapter is to explain this way of avoiding the problem of induction. Before turning to that task, some background must be provided.

In Section 1 I discuss the relationship between metaphysics and epistemology as it pertains to the problem of induction. In the present context, I assume a reliabilist theory of epistemic justification, and I focus exclusively on enumerative rather than non-enumerative induction. I outline these assumptions in Section 2. Section 3 presents the objection that the approach falls prey to

Howard Sankey, *Induction and Natural Kinds Revisited* In: *Reconsidering Causal Powers: Historical and Conceptual Perspectives*. Edited by: Benjamin Hill, Henrik Lagerlund, and Stathis Psillos, Oxford University Press (2021).
© Howard Sankey.
DOI: 10.1093/oso/9780198869528.003.0013

the charge of circularity. In Section 4 I outline the relevant aspects of Kornblith's account of induction. In Section 5 I briefly present Ellis's theory of kinds. In Section 6 I present my own view. In Section 7 I respond to the circularity objection. In Section 8 I consider an objection that may be raised against my response. Section 9 presents a brief conclusion.

1. A Metaphysical Starting Point

W. V. Quine's paper 'Natural Kinds' (1969) is an ancestor of the approach to induction that I propose. Quine argues on evolutionary grounds that our subjective similarity spacings reflect real divisions in nature. As he famously put the point, 'Creatures inveterately wrong in their inductions have a pathetic but praiseworthy tendency to die before reproducing their young' (126). Quine recognizes that his appeal to evolutionary considerations leads in a circle. It employs the inductively based theory of evolution to provide the rationale for induction. But he declines to take the charge of circularity seriously, remarking that there is 'no first philosophy' (127).

On one way of understanding first philosophy, epistemology is prior to metaphysics.[1] We must first establish that we are able to have knowledge. Only once this is done, may we proceed to the task of determining how the world actually is. This is a traditional thought in philosophy familiar from Descartes's *Meditations on First Philosophy*. Descartes seeks to determine what may be known with certainty. He arrives at the view that certainty requires clear and distinct ideas. Only then does he turn to the question of how to show that we have knowledge of the external world. Thus, for Descartes, epistemology is first philosophy. He seeks to show that we are able to have knowledge before turning to the nature and existence of the external world.

I propose an inversion of this Cartesian order. In at least one instance, metaphysics is prior to epistemology. We must first establish that something metaphysical is the case. Based on this, we may proceed to solve the epistemological problem.

I wish to argue that we should adopt a metaphysical stance as starting point before turning to the problem of induction.[2] The problem of induction is how

---

[1] One need not take epistemology to be first philosophy. Ellis (2009, 115–40) argues that the ontology of scientific realism may play the role of first philosophy.

[2] Michael Devitt's (2010) slogan 'put metaphysics first' is apposite here.

to provide a non-circular justification of induction: how may we justify induction without relying upon induction itself in the course of the justification? My suggestion is that it is the way the world is that makes induction reliable. The world contains natural kinds whose members have essential properties. Possession of essential properties by the members of natural kinds is what makes inductive inference reliable. The trick is to show that this appeal to the way the world is does not proceed in a circular manner.

2. Reliabilism and Enumerative Induction

I will assume a reliabilist account of epistemic justification. In seeking to justify induction, I aim only to provide an account of the reliability of induction. For present purposes, if a satisfactory account of the reliability of induction can be given, that suffices to justify induction. There is no need to go on to argue that we are justified to employ a reliable form of inference or belief-formation.[3]

To simplify matters, I focus on enumerative induction. For example, suppose we infer that all platypuses have webbed feet because all of the platypuses that we have observed have had webbed feet. This is an instance of enumerative induction. In such an inference, we generalize from a limited number of observed items to a claim about all members of the kind to which the items belong. My thesis is that such enumerative induction is grounded in natural kinds. It is because all members of the natural kind platypus have webbed feet that the inductive inference from observed to unobserved platypuses yields a correct conclusion.[4]

For present purposes, the question of the applicability of my account to more sophisticated forms of induction may be safely set to one side. This should not be taken to suggest that all induction is enumerative. Nor is it to

---

[3] Those of a more internalist persuasion may baulk at my use of the term 'justification' in the context of a generally reliabilist approach to induction. To partially assuage such qualms, I will simply mention Sosa's (1980, 7) suggestion that there is a further sense of 'justify' in addition to the idea of justification of a belief on the basis of a state with propositional content. An action may be justified by means of its consequences rather than on the basis of some propositional content. It is this latter sense of 'justified' that is apposite here.

[4] It may be objected that having webbed feet is not an essential property of duck-billed platypuses. If a platypus were born without webbed feet, it would still be a platypus. This may be granted without compromising the basic point. Rather than webbed feet, we may substitute instead a more suitable candidate for essential property of platypus, perhaps a particular genetic code. The induction about webbed feet would go through because having webbed feet is a property that depends upon the essential property.

suggest that non-enumerative forms of induction are unimportant. I seek to provide a justification for induction, not an account of all of our inductive practices. If the approach succeeds in the case of enumerative induction, it may plausibly be extended to other forms of induction. If the approach fails to do justice to enumerative induction, it is unlikely to provide suitable warrant for non-enumerative forms of induction.

This is simply a matter of taking things one step at a time. If enumerative induction may be justified in the way that I propose, then it is worth exploring the implications of the approach for less basic forms of induction. If the approach fails in the case of enumerative induction, there is little hope of extending it to other forms of induction.

3. The Charge of Circularity

'Induction and Natural Kinds' originally appeared in 1997. At the time, I was concerned that my approach might fall prey to the charge of circularity. Hence, I explicitly addressed the issue in the paper. A revised version of the paper was included in my book *Scientific Realism and the Rationality of Science* (Sankey 2008). In the revised version, I sought to improve the approach by drawing on work by David Papineau (1992). Though I remain sympathetic to Papineau's view, I now have a further proposal to make, which has been prompted by some critical comments by Stathis Psillos (2009).

In his review of *Scientific Realism and the Rationality of Science*, Psillos presents a number of objections to my approach to induction. The chief objection is that the approach employs ampliative inference to justify ampliative inference, and thereby fails to avoid circularity. Psillos writes as follows:

> Sankey's gambit is smart: it assumes the existence of natural kinds to *explain* the reliability of induction. Should it work, it promises to avoid the well-known charge of circularity. The move from the success of scientific methodology to the existence of natural kinds is abductive: the existence of natural kinds is taken to be the best explanation of the success of science. If so, the existence of natural kinds can do apparently non-circular work in justifying induction.... it is wrong to believe that IBE can bypass the problem of induction—since the problem concerns, at bottom, the very idea of an *ampliative* but rational method. IBE has no atemporal warrant—as Hume in effect observed when he criticized a standard appeal to active

powers to justify induction....Nor is it obvious that essential properties explain the reliability of induction in a non-circular way, since, as Sankey himself notes, 'good inductive inferences project essential properties, whereas bad ones project accidental properties.' Unless there is an independent way to classify inductions into good and bad ones, essentialism cannot ground the reliability of induction.   (682)

In this passage, Psillos raises a number of critical points against my position. I will focus on what I understand to be the central objection. I take the central objection to be that my approach fails to avoid the charge of circularity.

In the revised version of 'Induction and Natural Kinds', I sought to forestall the circularity objection by adopting an approach with which Psillos (1999, 82) appears to be sympathetic. This is the approach adopted by David Papineau (1992) in his inductive defence of the reliability of induction. Papineau argues by induction from the past reliability of induction to its general reliability. Against the charge of circularity, Papineau appeals to the distinction between premise and rule circularity. A premise-circular argument contains the conclusion itself in the premises. A rule-circular argument uses a rule of inference (for example induction, deduction) to support the rule itself. The conclusion that induction is reliable does not appear as a premise in the argument for the reliability of induction. Papineau's inductivist defence of induction is not therefore premise-circular. But it is rule-circular.

I continue to regard the idea of rule circularity as an important tool to be employed in the inductive justification of induction. But I now wish to propose another way to avoid the charge of circularity. My goal here is to explain this other way. Before I turn to that, let me describe the approach to induction in greater detail. The approach is not new. Nor is it mine alone.

4. Hilary Kornblith on Induction

My approach to the problem of induction is related to the venerable idea that our use of induction rests upon a principle of the uniformity of nature. In my view, induction is reliable because the world makes it so. In effect, I endorse a version of the principle of uniformity according to which induction is justified as a result of nature being uniform. This principle is not something that we know a priori to be true. Nor is it to be thought of as a logical presupposition of inductive inference. The principle is an empirical claim about the nature of reality. It is an a posteriori metaphysical claim.

I draw explicitly on the position developed by Hilary Kornblith (1993). Kornblith presents an account of the reliability of inductive inference which rests on two key claims. On the one hand, the world has a natural-kind structure. That is, the world contains individual items which are members of natural kinds. On the other hand, our minds reflect the natural-kind structure of the world. We conceptualize the world in a way that is sensitive to its natural-kind structure. Moreover, our inductive inferences presuppose the existence of such a structure. Given the way our minds reflect the kind structure of the world, there is a 'dovetail fit' between mind and the world. This dovetail fit explains the reliability of inductive inference.

In my account of induction, I emphasize the role of natural kinds in explaining the reliability of induction. I focus primarily on Kornblith's view that natural kinds provide the ground for induction rather than on his account of how the human mind fits with the natural-kind structure of the world. But Kornblith is surely right that an account must be given of how we conceptually grasp natural kinds and reason inductively in a manner that is informed by the natural-kind structure of the world. I do not have anything substantive to say about this aspect of the position. I shall simply assume that an approach along the lines developed by Kornblith is well motivated and in broad terms correct.

What sort of argument does Kornblith provide for the approach? Kornblith argues that the natural-kind structure of the world is the best explanation of the reliability of induction. His use of this form of argument is especially apparent when he considers the role played by natural kinds in inductive inference in the sciences. In that context, his argument takes a form that is familiar from the literature on scientific realism:

> If the scientific categories of mature sciences did not correspond, at least approximately to real kinds in nature, but instead merely grouped objects together on the basis of salient observable properties which somehow answer to our interests, it would be utterly miraculous that inductions using these scientific categories tend to issue in accurate predictions. Inductive inferences can only work, short of divine intervention, if there is something in nature binding together the properties which we use to identify kinds. Our inductive inferences in science have worked remarkably well, and, moreover, we have succeeded in identifying the ways in which the observable properties which draw kinds to our attention are bound together in nature. In light of these successes, we can hardly go on to doubt the existence of the very kinds which serve to explain how such successes were even possible.   (41–2)

In this passage, Kornblith's inference to the best explanation of reliable inductive inference in science bears a clear and striking resemblance to the 'no miracles' argument for scientific realism made famous by Hilary Putnam. Given this reliance on an ampliative form of inference, however, Kornblith's approach would appear to be vulnerable to the same charge of circularity as is my own version of the approach.[5]

As indicated, Kornblith's account of induction rests on the co-occurrence of properties among members of a natural kind. In order to explain such co-occurrence of properties, Kornblith adopts the homeostatic property cluster (HPC) account of natural kinds proposed by Richard Boyd (1989; 1991). On Boyd's account, natural kinds possess groups of properties which enter into a relation of homeostatic equilibrium: 'A natural kind is a cluster of properties which, when realized together in the same substance, work to maintain and reinforce each other, even in the face of changes in the environment' (Kornblith 1993, 35). On the HPC account, only certain groups of properties are able to co-occur in stable assemblages. It is this fact about the way in which properties may be arranged in only a limited number of ways that provides the basis for reliable inductive inference: 'Because there are natural kinds, and thus clusters of properties which reside in homeostatic relationships, we may reliably infer the presence of some of these properties from the presence of others. In short, natural kinds make reliable inductive inference possible, because were it not for the existence of these homeostatic clusters, the presence of any set of properties would be fully compatible with the presence of any other' (36). For Kornblith, the way in which certain groups of properties are able to reside together in homeostatic equilibrium is what provides the metaphysical underpinning of inductive inference. This is why he thinks of natural kinds as providing 'the natural ground of inductive inference' (36).

No doubt, the HPC account may be fruitfully applied to natural kinds within the life sciences. It is less clear that it may be extended to non-biological kinds. The account rests on a metaphor that applies in the case of living organisms. It does not readily generalize to physical entities other than organisms. For this reason, I choose not to follow Kornblith in adopting Boyd's account of natural kinds.

---

[5] As Kornblith does not address the charge of circularity, it is difficult to determine how he might respond to it. It is entirely possible that his attitude toward the charge might parallel Quine's naturalistic dismissal of the issue.

5. Brian Ellis on Natural Kinds

My own preference is to adopt an essentialist theory of natural kinds of the kind proposed by Brian Ellis (2001). Such an account seems to me to provide a more solid foundation for inductive inference than does the HPC account.[6]

According to Ellis, individual entities such as electrons belong to natural kinds. But individual entities are not the only things that belong to natural kinds. In addition to natural kinds of entities or objects, there are also properties such as charge and mass, as well as relations such as gravitational attraction or chemical bonding. Moreover, there are natural kinds of events and processes, such as chemical reactions or radioactive decay.

The world in which we live is highly structured. This is largely due to the fact that it is divided up into categorically distinct natural kinds. Natural kinds have objectively existing boundaries which we discover. They do not overlap or fade into each other. The borderline between them is not an arbitrary or blurred one that depends on a human classificatory decision or convention. The only way that kinds may overlap is by way of a relation of inclusion such as the relation of genus to species. The system of natural kinds forms a hierarchical structure. More specific kinds belong to more general kinds. General kinds possess essential properties which are also found in more specific kinds. The specific kinds within a general kind all possess the essential properties of the general kind.

For Ellis (1999, 19), the essential properties of natural kinds are intrinsic properties shared by all members of the same kind. The essential properties of natural kinds of things are dispositional. They have 'the nature of powers, capacities and propensities'. Natural kinds are characterized by the intrinsic causal powers of the things that belong to those kinds. This permits a connection to be drawn between laws of nature and natural kinds. Laws of nature describe the essences of natural kinds. They are statements that are made true by the causal powers of things that belong to natural kinds. The truthmakers for laws of nature are the causal powers or intrinsic dispositions of natural kinds of things. Because the causal powers or dispositions are essential properties of natural kinds, the laws of nature are not contingent.

---

[6] I do not wish to rule out the possibility that a position like the HPC account might be combined with the essentialist view of Ellis's that we are about to consider. The HPC account seems to be designed for biological rather than physical kinds. By contrast, Ellis's (2002, 31) account is designed primarily for physical and chemical kinds. Ellis considers biological kinds to be clusters of micro-species, which suggests there might be room for reconciliation with the Boydian view.

They are metaphysically necessary, grounded in the essential properties of natural kinds of things.

6. My Approach

On the view that I propose, induction is justified because nature is uniform. There is a sense, as we have seen, in which my approach involves a principle of the uniformity of nature. But the principle of uniformity is not to be understood in the way such principles are usually understood. It is not an overarching or blanket statement to the effect that the future resembles the past. Nor is it simply a global statement that nature is uniform.

The principle relates specifically to the essential properties of natural kinds. On this version of the principle, nature is uniform in the precise sense that there are natural kinds whose members all possess a shared set of essential properties. In light of the relation that obtains between natural kinds and laws of nature, to say that nature is uniform is to say that it is governed by laws of nature. On the essentialist view that I take over from Ellis (2001), laws of nature are grounded in the causal powers that are the essential properties of natural kinds of things.

What I wish to suggest is as follows: when we use induction to form a belief about the future, we are justified in doing so because nature is in fact uniform. It is uniform in the sense that there are natural kinds of things which possess sets of essential properties. All members of a kind have the same essential properties. Unobserved members of a kind possess the same essential properties as observed members of the kind. The fact that observed and unobserved members of a kind possess the same essential properties is what makes induction reliable. When we predict that an unobserved object will have an essential property that observed objects of the same kind have, our prediction will be correct. It will be correct because all members of the kind have essential properties in common. It is because all members of the kind have the property in common that our prediction is correct.

Of course, the conclusion of an inductive argument need not be correct. On the present account, one way in which an inductive inference may lead to a false conclusion is for essential properties to be mistakenly identified as such. If a non-essential property were mistakenly taken to be an essential property, future unobserved members of the kind may fail to have the property mistakenly thought to be essential to the kind. In such a situation, the prediction that a future unobserved member of the kind will possess the property in question may be mistaken.

As so far stated, the account requires two qualifications. The first relates to the apparent restriction of inductive inference to essential properties of natural kinds. The second relates to the question of how inductive inference may be employed with respect to items that do not belong to natural kinds.

With regard to the first point, I have presented the account in terms of the essential properties of natural kinds. But this is an oversimplification. Inductive inference need not be restricted to essential properties. There may be non-essential properties that depend on essential properties in a systematic way. Where there are non-essential properties that depend in a systematic way on essential properties, it may be possible for there to be reliable inductive inference that ranges over such dependent non-essential properties.

To take a particular example, being black is unlikely to be an essential property of ravens. In the case of ravens, the property of being black depends in some systematic way on facts about the genetic make-up of ravens that typically give rise to black pigmentation. The essential properties of ravens will likely be found within the genetic code of ravens.[7] Hence, the reliability of an inductive inference about the blackness of ravens is not due to the fact that blackness is an essential property of ravens. It is due to the fact that the blackness of ravens depends in some systematic way on features of the genetic code of ravens that constitute the essential properties of ravens.

Second, as so far presented, my account of induction applies to inductive inference about members of natural kinds. But we routinely infer by induction about members of non-natural kinds, such as trains, trams, and aeroplanes. Surely, inductive inference is not restricted to the essential properties of natural kinds and the non-essential properties that systematically depend upon the essential properties of natural kinds. We are perfectly able to reason by induction about items that do not belong to natural kinds.[8]

In the original development of the approach, I was inclined to adopt an uncompromising stance with respect to this issue. I initially took the view that induction about entities that do not belong to a natural kind may only be reliable to the extent that the entities are composed of parts that do belong to natural kinds. It is only in virtue of the fact that the component parts consist of material belonging to natural kinds that inference about members of non-natural kinds may be reliable. I found encouragement for this stance in a

---

[7] A qualification may be in order. Arguably, being a bird is essential to being a raven. Having a beak, feathers, and wings is essential to being a bird. This may or may not reside in some bit of genetic code common to all birds.

[8] The objection may be put another way. Induction is a non-deductive form of inference in which the premises provide support for the conclusion but do not deductively entail the truth of the conclusion. Nothing about the definition of induction implies that induction must be restricted to natural kinds.

remark by Wilkerson (1995, 32): 'because there are no very specific real essences that make rubbish rubbish, and tables tables, I cannot even in principle make sound inductive projections about rubbish as such or tables as such.' Wilkerson's suggestion was that only natural-kind predicates are projectible and thus able to be employed in inductive inference.[9] This seemed, in turn, to suggest that reliable induction might only be grounded in natural kinds, since it is only the regularity built into natural kinds that may provide the basis for such reliability.

The problem with such an uncompromising stance is that there are cases of apparently sound inductive inference not underpinned by the existence of a natural kind. Apart from cases in which we infer inductively about items that belong to non-natural kinds, we may also make legitimate inductive inferences about singular items considered simply as individual things rather than as instances of a kind.[10] Because there are cases of reliable induction for which the existence of natural kinds does not appear to be responsible, a less uncompromising stance may be appropriate. Hence, I wish only to assert that my account of induction is able to explain the reliability of an important and presumably large class of cases of inductive inference, namely, those in which induction applies to members of natural kinds, as well as to items consisting of natural kinds which are responsible for the regularity picked out by the induction. It remains to be explained how and why induction is reliable in cases where the presence of natural kinds plays no apparent role.

7. The Circularity Objection Again

I now return to the problem that this chapter seeks to address. As we have seen, my aim is to provide an account of the reliability of inductive inference and thereby to justify induction. Does this approach avoid the problem of induction? According to Psillos (2009), my approach employs an ampliative inference to argue in support of an ampliative inference. If this is right, the approach seems not to avoid the charge of circularity at all.

---

[9] Wilkerson's mention of projectibility brings out another attractive feature of the present approach, namely, that it promises to resolve Goodman's problem of the grue emeralds. But it may not all be plain sailing. As Bruce Langtry has pointed out to me, 'grue' is defined in terms of 'green' and 'blue', which may be natural-kind predicates. So, it is not immediately clear how appeal to natural-kind terms does solve the grue problem.

[10] For example, we may infer from the fact that a particular potted plant has thrived after being watered that it always thrives after being watered. Such an inductive inference need make no reference to the kind to which the plant belongs and indeed may be restricted specifically to the one plant under consideration. (I owe this example to Greg Restall.)

Let me state the objection more precisely. The argument for the existence of natural kinds proceeds by inference to best explanation (IBE). A similar argument is employed to support the claim that induction is reliable because it is grounded in natural kinds. But IBE and induction are both forms of ampliative inference. So, an ampliative inference is employed to justify an ampliative inference. The circularity is immediately apparent.

As mentioned previously, I earlier sought to forestall this objection by appeal to Papineau's distinction between rule and premise circularity. Though I continue to hold that this distinction has a role to play in a reliabilist justification of induction, I now favour another response to the circularity objection. Put simply, the response is that the IBE is not what does the justificatory work in the justification of induction. It is the natural kinds that provide the justification for induction.

To develop the point in more detail, it is important to reflect upon the role played by IBE in the account. There are two places where IBE is employed. First, an IBE is employed to argue for the existence of natural kinds. Second, an IBE is used to argue that the existence of real kinds in nature is what makes induction reliable. But notice that on this account the reliability of induction does not itself depend upon the IBE. What underpins the reliability of induction is not the IBE. It is the natural kind structure of the world that makes induction reliable. IBE has nothing to do with it.

To see this, suppose that we correctly employ induction to predict that a previously unobserved member of a kind will in future be found to have some specific property. On my account, the explanation of why the induction leads to a correct prediction is that it correctly picks out a real pattern in nature. Thus, it is the way the world is that makes the inference reliable. It is not the inference to the best explanation of the reliability of induction that makes the induction itself reliable. The reliability of the inductive inference does not depend upon the IBE.

The point may be presented in another way. It may be illustrated in terms of a once popular distinction in the philosophy of science. Karl Popper (1959) and Hans Reichenbach (1938) drew a distinction between the context of discovery and the context of justification. They distinguished between the way in which one arrives at a theory (the context of discovery) and the way in which the theory is justified (the context of justification). The initial inspiration or creative insight that gives rise to a theory need have nothing to do with the tests or experiments which provide empirical evidence on the basis of which the theory is to be accepted.

A similar point applies to the present approach to induction. IBE is employed to argue that the world has a natural-kind structure, and that this structure underlies the reliability of inductive inference. This is the context of discovery. We employ IBE in the context of discovery to argue for an account of the natural ground of induction. But our use of IBE to argue for the account is not what provides the ground for induction. What justifies our use of induction is the fact that the world is a certain way. It is the natural kind structure of the world that makes induction reliable. In short, IBE is how we discover the ground of induction. It is not what grounds the induction.[11]

In sum, I propose an argument to the effect that induction is reliable, and so justified, because the world has a natural-kind structure that makes induction reliable. If I am right, the approach is not circular. It is not circular because the IBE is not what grounds induction. It is the world that grounds induction. The only role played by IBE is to discover that natural kinds are what make induction reliable.

8. A Further Objection

In the previous section, I attempted to meet the charge of circularity by arguing that the justification of induction rests on the natural kind structure of the world rather than upon the IBE given on behalf of this account of induction. I will now consider an objection that might be raised against this response to the charge of circularity. The objection might be presented in the following terms: granted, if natural kinds exist, induction will be reliable. But we must use IBE to arrive at a justified belief in the existence of natural kinds. Moreover, we must use IBE to justify the belief that the reliability of induction is based on natural kinds. This means that the justification of induction depends upon an IBE about the reliability of induction. So, in the end, you use ampliative inference to justify induction.

On the view that I have proposed in this chapter, induction is justified because the existence of natural kinds makes inductive inference reliable. As I have pointed out, an IBE is employed to argue for this view of the warrant of induction. But I wish to deny that the IBE employed in the argument plays the role in justifying induction that the objection assumes it to play.

---

[11] I do not think that the analogy between the two contexts distinction and the present approach to induction is a perfect analogy. But the distinction between the two contexts helps to illustrate the distinction between, on the one hand, the argument for the account of reliable induction and, on the other hand, the natural-kind structure that underlies the reliability of induction, which thereby provides the warrant for induction.

The objection is based on an internalist assumption that we must form a justified belief in the reliability of induction in order to be justified in our use of induction.[12] But I do not see why we must form such a belief in order to be justified in the use of induction. The existence of natural kinds would render our use of induction reliable even if we did not perform the IBE and thereby come to hold the present view of induction. In order for use of induction to be justified, there is no need for us to know or to justifiably believe that induction is warranted by the existence of natural kinds.

To employ the above internalist assumption against the position that I have presented begs the question against the reliabilist stance adopted here. But, quite apart from that, the objection relies upon a contentious view about what is required for justification. The objection relies on an assumption similar in spirit to the infamous KK-thesis that in order to know we must know that we know. In particular, it assumes that in order be justified in our use of induction we must hold a justified belief that induction is justified. This in turn assumes that in order to be justified in a belief we must justifiably believe that we are justified in the belief.

It would be wrong to reject such an assumption out of hand. At the same time, it would be wrong to adopt the assumption without strong positive arguments in its favour. More generally, the assumption is so fundamentally at odds with the approach adopted here that it is more than just an objection. It is a wholesale rejection of the approach to the problem of induction that I have sought to sketch in this chapter.

9. Conclusion

In this chapter, I have attempted to respond to the charge of circularity raised by Stathis Psillos (2009) against my reliabilist justification of induction. Against the charge that my account employed ampliative inference to justify an ampliative inference, I have argued that what grounds induction is not the IBE presented for the reliabilist account of induction. It is the natural kind structure of the world that makes induction reliable. Given this, the circularity may be avoided.

---

[12] I interpret the notion of justification employed in the objection in internalist terms. However, the notion of justification could be interpreted in reliabilist terms. If so, it is not clear that the objection poses a serious threat to the approach. Nor, however, is it clear that the assumption on which the objection is based need be granted even if justification is understood in reliabilist terms.

INDUCTION AND NATURAL KINDS REVISITED

I conclude in programmatic terms. As indicated, I have presented the position as an account of the reliability of enumerative induction. If the account can be sustained, then it may be worthwhile exploring the prospects of extending the approach to non-enumerative forms of induction. We have also seen that there appear to be cases of inductive inference which do not apply to natural kinds. It remains to be seen whether the account may be extended to cases where induction seems not to be applied to natural kinds or whether some alternative approach must be developed for cases in which induction does not apply to natural kinds. These are topics for future work which must be addressed if the present approach is to be developed in greater detail.

References

Boyd, Richard. 1989. 'What realism implies and what it does not'. Dialectica, 43, nos. 1–2 (June): 5–29.

Boyd, Richard. 1991. 'Realism, anti-foundationalism and the enthusiasm for natural kinds'. *Philosophical Studies* 61, nos. 1–2 (February): 127–48.

Devitt, Michael. 2010. *Putting Metaphysics First: Essays on Metaphysics and Epistemology*. Oxford: Oxford University Press.

Ellis, Brian. 1999. 'Causal Powers and Laws of Nature'. In *Causation and Laws of Nature*, ed. Howard Sankey, Australasian Studies in History and Philosophy of Science 14, 19–34. Dordrecht: Springer.

Ellis, Brian. 2001. *Scientific Essentialism*. Cambridge: Cambridge University Press.

Ellis, Brian. 2002. *The Philosophy of Nature: A Guide to the New Essentialism*. Chesham: Acumen Press.

Ellis, Brian. 2009. *The Metaphysics of Scientific Realism*. Durham: Acumen Press.

Kornblith, Hillary. 1993. *Inductive Inference and its Natural Ground*. Cambridge: MIT Press.

Papineau, David. 1992. 'Reliabilism, Induction and Scepticism'. *Philosophical Quarterly* 42, no. 166 (January): 1–20.

Popper, Karl. 1959. *The Logic of Scientific Discovery*. New York: Basic Books

Psillos, Stathis. 1999. *Scientific Realism: How Science Tracks Truth*. New York: Routledge.

Psillos, Stathis. 2009. Review of *Scientific Realism and the Rationality of Science*, by Howard Sankey. *Australasian Journal of Philosophy* 87 (4):681–84.

Quine, W. V. 1969. 'Natural Kinds'. In *Ontological Relativity and Other Essays*, 114–38. New York: Columbia University Press.

Reichenbach, Hans. 1938. *Experience and Prediction: An Analysis of the Foundations and Structure of Knowledge*. Chicago: University of Chicago Press.

Sankey, Howard. 1997. 'Induction and Natural Kinds'. *Principia* 1 (2): 239–54.

Sankey, Howard. 2008. *Scientific Realism and the Rationality of Science*. Aldershot: Ashgate.

Sosa, Ernest. 1980. 'The Raft and the Pyramid: Coherence versus Foundations in the Theory of Knowledge'. In 'Studies in Epistemology', *Midwest Studies in Philosophy* 5: 3–26.

Wilkerson, T. E. 1995. *Natural Kinds*. Aldershot: Avebury.

Index